Column Chromatography
http://dx.doi.org/10.5772/47823
Edited by Dean F. Martin and Barbara B. Martin

Contributors
Sylwester Czaplicki, Zhang Xiaopo, Dean Frederick Martin, Yasser Moustafa, Rania Morsi, Alaíde S. Barreto, Gláucio Diré Feliciano, Özlem Bahadır Acıkara, Gulcin Saltan Citoglu, Serkan Özbilgin, Burçin Ergene, Ana Cláudia F. Amaral

Published by InTech
Janeza Trdine 9, 51000 Rijeka, Croatia

Publishing Process Manager Viktorija Zgela
Technical Editor InTech DTP team
Cover InTech Design team

Additional hard copies can be obtained from orders@intechopen.com

Column Chromatography, Edited by Dean F. Martin and Barbara B. Martin
p. cm.
ISBN 978-953-51-1074-3

For EU product safety concerns:
IN TECH d.o.o., Prolaz Marije Krucifikse Kozulić 3, 51000 Rijeka, Croatia,
info@intechopen.com or visit our website at intechopen.com.

Meet the editors

Dean F. Martin, Ph.D. is a distinguished university professor of chemistry emeritus at the University of South Florida, where he joined the faculty in 1964, attaining emeritus status in 2006. Trained as an inorganic chemist, his research interests have been concerned with applications of coordination chemistry to a range of problems including use of ion-exchange chromatography in the synthesis of organothallium perchlorates, red tide outbreaks, management of nuisance aquatic plants, and removal of nuisance anions and pharmaceuticals from the environment. He is the author or co-author of over three hundred articles in refereed journals, five books and has over 25 years of editorial experience.

Barbara B. Martin, M.S. is an assistant professor of chemistry (courtesy) at the University of South Florida. She graduated from Grinnell College (B.A., 1956) and The Pennsylvania State University (M.S., 1959). She is the co-author of the book, Coordination Compounds, and 72 research articles in refereed journals. She was a co-recipient of the Florida Medal of the Florida Academy of Sciences. She served as a Co-Editor of Florida Scientist for 26 years and became the first Editor Emerita in 2010.

Contents

Preface

As Moustafa and Morsi remind us in Chapter One, it has been just over 100 years since a Russian botanist, M. Tswett discovered chromatography. It seems timely to review the current status of the field of chromatography and appreciate the many improvements that have been made in the field. To do this thoroughly, of course, would require many volumes, but within the limits of space, the chapters that follow surely do indicate the range of techniques and some of the important applications. It is to be hoped that careful readers will reflect on these and consider other applications.

Accordingly, Chapters Two through Five look at different techniques of chromatography with a consideration of applications. These include ion-exchange chromatography (Chapter Two), the use of chromatography to characterize the bioactivity of compounds (Chapter Three), affinity chromatography and the utilization in drug discovery (Chapter Four), and the use of column chromatography with chelating agents attached to useful substrates (Chapter Five).

Finally, this volume provides three examples of the range of utilization of chromatography in the study of natural products. This section provides useful and, it is hoped, inspirational examples of how far the field of chromatography has come with respect to natural products since Tswett's discovery. These examples are considered in Chapters Six, Seven, and Eight.

This book is characterized by three important features. The authors represent an impressive collection of international workers from Brazil, China, Egypt, Poland, Turkey, and the United States. The majority of the chapters reflect the importance of collaborative efforts in contemporary research. Finally, some chapters are especially useful because of the experimental details that are provided.

And it is to be hoped that readers will find that the chapters are both informative and inspirational.

Dean F. Martin
Barbara B. Martin

Chapter 1

Ion Exchange Chromatography - An Overview

Yasser M. Moustafa and Rania E. Morsi

Additional information is available at the end of the chapter

http://dx.doi.org/10.5772/55652

1. Introduction

Chromatography is the separation of a mixture of compounds into its individual components based on their relative interactions with an inert matrix. However, chromatography is more than a simple technique, it is an important part of science encompassing chemistry, physical chemistry, chemical engineering, biochemistry and cutting through different fields. It is worth to be mentioned here that the IUPAC definition of chromatography is "separation of sample components after their distribution between two phases".

1.1. Discovery and history of chromatography [1, 2]

M. Tswett (1872-1919), a Russian botanist, discovered chromatography in 1901 during his research on plant pigments. According to M. Tswett: "An essential condition for all fruitful research is to have at one's disposal a satisfactory technique". He discovered that he could separate colored leaf pigments by passing a solution through a column packed with adsorbent particles. Since the pigments separated into distinctly colored bands as represented in Figure 1, he named the new method "chromatography" (chroma – color, graphy –writing). Tswett emphasized later that colorless substances can also be separated using the same principle.

The separation results from the differential migration of the compounds contained in a mobile phase through a column uniformly packed with the stationary matrix. A mobile phase, usually a liquid or gas, is used to transport the analytes through the stationary phase while the matrix, or stationary phase, is generally an inert solid or gel and may be associated with various moieties, which interact with the analyte(s) of interest. Interactions between the analytes and stationary phase are non-covalent and can be either ionic or non-ionic in nature depending on the type of chromatography being used. Components exhibiting fewer interactions with the stationary phase pass through the column more quickly than those that interact to a greater degree.

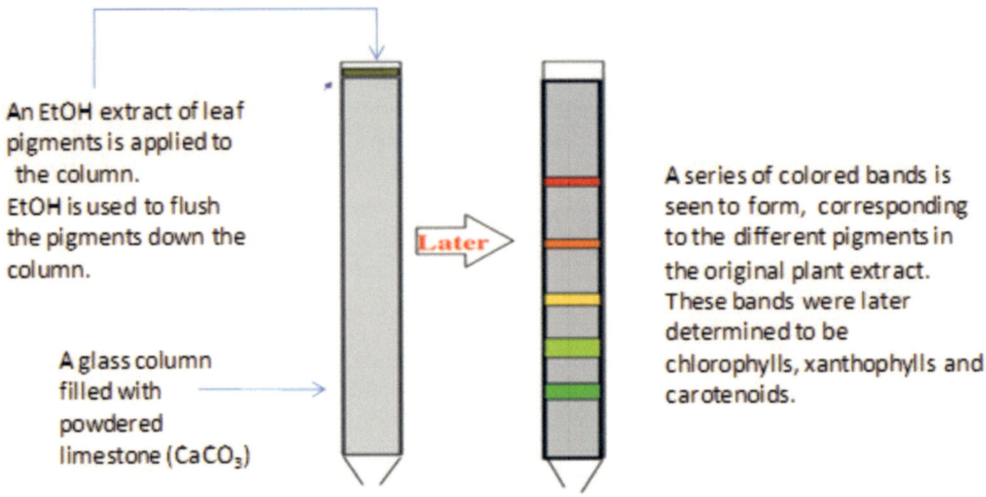

Figure 1. Schematic diagram of the principles of chromatography as discovered by Tswett (1901).

Tswett's initial experiments involved direct visual detection and did not require a means of quantitation. Nowadays, chromatography is not only a separation technique. In most versions, it is hyphenated analytical techniques combining the separation with the identification and quantitative determination of the separated components. In this form, chromatography has become the most widely used technique in the chemical analysis of complex mixtures.

Many versions of chromatography are used. The various chromatographic techniques are subdivided according to the physical state of these two phases, the mobile and the stationary phases. These are: liquid chromatography including high performance, ion, micellar, electrokinetic, thin-layer, gel-permeation, and countercurrent versions; gas chromatography and supercritical fluid chromatography. Various forms of chromatography can be used to separate a wide variety of compounds, from single elements to large molecular complexes. By altering the qualities of the stationary phase and/or the mobile phase, it is possible to separate compounds based on various physiochemical characteristics. Among these characteristics are size, polarity, ionic strength, and affinity to other compounds. Chromatography also permits a great flexibility in the technique itself. The flow of the mobile phase might be controlled by gravity, pressure, capillary action and electro-osmosis; the separation may be carried out over a wide temperature range and sample size can vary from a few atoms to many kilograms. Also, the shape of the system in which the separation takes place can be varied, using columns of various length and diameter or flat plates. Through all this, evaluation chromatography has been transformed from an essentially batch technique into an automated instrumental method. Through its continuous growth, chromatography became the most widely used analytical separation technique in chemistry and biochemistry. Thus, it is not exaggeration to call it the technique of the 20th Century.

http://dx.doi.org/10.5772/55652

2. Ion chromatography

Classical liquid chromatography based on adsorption- desorption was essentially a non-linear process where the time of retardation (retention time) and the quantitative response depend on the position on the adsorption isotherm. Essentially, it was a preparative technique: the aim was to obtain the components present in the sample in pure form which could then be submitted to further chemical or physical manipulations [3].

Ion exchange chromatography (or ion chromatography, IC) is a subset of liquid chromatography which is a process that allows the separation of ions and polar molecules based on their charge. Similar to liquid chromatography, ion chromatography utilizes a liquid mobile phase, a separation column and a detector to measure the species eluted from the column. Ion-exchange chromatography can be applied to the determination of ionic solutes, such as inorganic anions, cations, transition metals, and low molecular weight organic acids and bases. It can also be used for almost all kinds of charged molecule including large proteins, small nucleotides and amino acids. The IC technique is frequently used for the identification and quantification of ions in various matrices.

2.1. Ion chromatography process [4]

The basic process of chromatography using ion exchange can be represented in 5 steps (assuming a sample contains two analytes A & B): eluent loading, sample injection, separation of sample, elution of analyte A, and elution of analyte B, shown and explained below. Elution is the process where the compound of interest is moved through the column. This happens because the eluent, the solution used as the solvent in chromatography, is constantly pumped through the column. The representative schemes below are for an anion exchange process. (Eluent ion = , Ion A= , Ion B = )

Step 1: The eluent loaded onto the column displaces any anions bonded to the resin and saturates the resin surface with the eluent anion.

This process of the eluent ion (E^-) displacing an anion (X^-) bonded to the resin can be expressed by the following chemical interaction:

$$Resin^+\text{-}X^- + E^- \Leftrightarrow Resin^+\text{-}E^- + X^-$$

Step 2: A sample containing anion A and anion B are injected onto the column. This sample could contain many different ions, but for simplicity this example uses just two different ions as analytes in the sample.

Step 3: After the sample has been injected, the continued addition of eluent causes a flow through the column. As the sample elutes (or moves through the column), anion A and anion B adhere to the column surface differently. The sample zones move through the column as eluent gradually displaces the analytes.

Step 4: As the eluent continues to be added, the anion A moves through the column in a band and ultimately is eluted first.

This process can be represented by the chemical interaction showing the displacement of the bound anion (A^-) by the eluent anion (E^-).

$Resin^+$-A^- + E^- <=> $Resin^+$- E^- + A^-

Step 5: The eluent displaces anion B, and anion B is eluted off the column.

$Resin^+$-B^- + E- <=> $Resin^+$-E- + B^-

The overall 5 step process can be represented pictorially as shown in Figure 2:

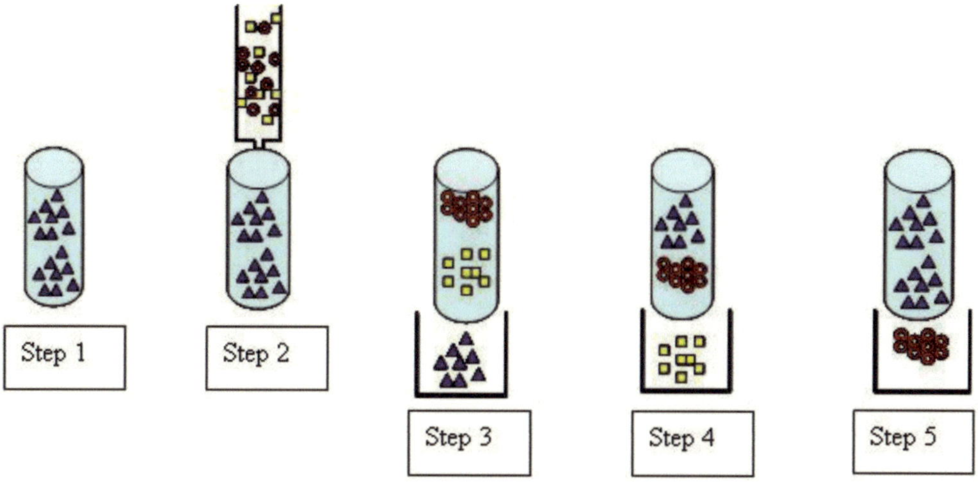

Figure 2. Schematic representation of IC process.

A typical ion chromatography consists of several components as shown in Figure 3. The eluent is delivered to the system using a high-pressure pump. The sample is introduced then flows through the guard and into the analytical ion-exchange columns where the ion-exchange separation occurs. After separation, the suppressor reduces the conductivity of the eluent and increases the conductivity of the analytes so they are delivered to the detector. A computer and software are used to control the system, acquire and process the data. Since the introduction of ion chromatography in 1975, many developments were carried out to improve suppressor technology to provide better sensitivity and consistency for the analysis of a wide variety of compounds [5].

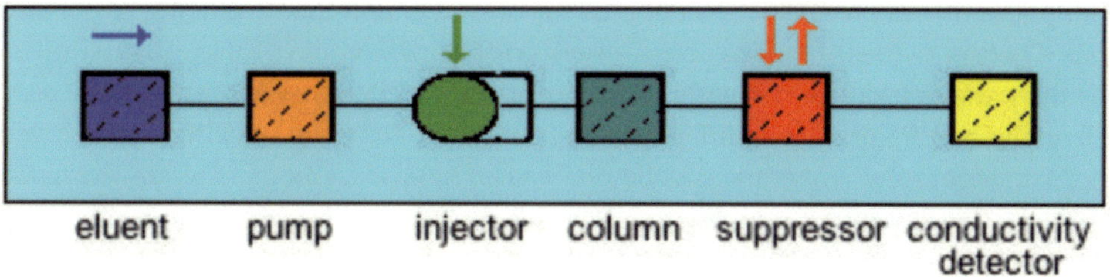

Figure 3. Schematic representation of Ion chromatography instrumentation.

3. Instrumentation [6-9]

Typical IC instrumentation includes: pump, injector, column, suppressor, detector and recorder or data system as represented in Figure 4.

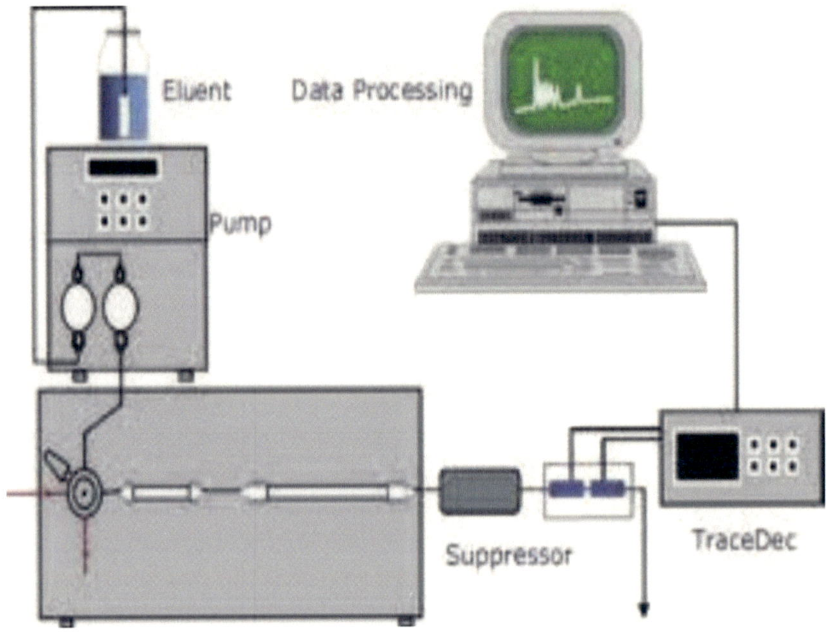

Figure 4. Typical ion chromatography instrument.

3.1. Pump

The IC pump is considered to be one of the most important components in the system which has to provide a continuous constant flow of the eluent through the IC injector, column, and detector. The most practical system for the delivery of the mobile phase is that which can combine several liquids in different proportions at the command of the operator. This blending capability speeds the process of selecting the optimum eluent mixture required for isocratic analysis. There is a series of mobile phase reservoirs that can contain a range of different mobile phases that can be used individually, blended or for mobile phase programming purposes "gradient elution". In general liquid chromatography, the reservoirs can be stainless steel but in ion chromatography where the mobile phases can have extreme pH values, the reservoirs need to be made of glass or preferably a suitable plastic such as PEEK (polyether-ether-ketone). The advantage of PEEK is that it is also inert to many organic solvents that may need to be used in the mobile phase. In fact, all components of an ion chromatograph that may come in contact with either phase of the distribution system should be constructed from appropriate inert material. This includes all mobile phase conduits, valves, pumps, sampling devices, columns, detector sensor cells, etc. The solvent reservoirs are connected to a solvent selection valve and a solvent programmer where a particular solvent or particular solvent program can be selected. The solvent then passes from the selector/programmer to a high pressure pump. The mobile phase passes from the pump to the sampling device, usually a simple rotating valve that on rotation places the sample in line with the mobile flow which then passes onto the column. The exit flow from the column passes either to an ion suppressor (which will be discussed later) or directly to the detector. Gas may come out of the solution at the column exit or in the detector, resulting in sharp spikes. Spikes are created by microscopic bubbles which change the nature of the flowing stream making it heterogeneous. The drift may occur as these microscopic bubbles gradually collected and combined in the detector cell. The best results can be obtained by applying vacuum to each solvent for about 5 min. with subsequent helium purging and storing under helium atmosphere.

3.1.1. Pumps types

The constant-flow pumps is the most widely used in all common IC applications. Flow rate stability is an important pump feature that distinguishes pumps. For size exclusion chromatography, the flow rate has to be extremely stable. External electronic control is a very desirable feature when automation or electronically controlled gradients are to be run.

3.1.2. Constant flow pumps

Constant-flow systems are generally of two basic types: reciprocating piston and positive displacement (syringe) pumps. Reciprocating piston pump can maintain a liquid flow for indefinitely long time.

3.1.3. Reciprocating piston pumps

The pumping rate is controlled by piston retracts or by the cam rotating speed. The main drawback of this type of pump is sinusoidal pressure pulsations which lead to the necessity of using pulse dampers.

3.1.4. Dual piston pumps

Provides a constant and almost pulse free flow. Both pump chambers are driven by the same motor through a common eccentric cam; this common drive allows one piston to pump while the other is refilling. As a result, the two flow-profiles overlap each other significantly reducing the pulsation downstream of the pump; this is visualized below.

Its advantages are: unlimited solvent reservoir allowing long-term unattended use; quick changeover and clean out capability; wide flow rate range (0.01 to 10 ml/min) is provided without gear change. While its drawbacks are: incompletely compensated pulsations might be observable at high refractive index detector sensitivities, especially at low flow rates; pump reliability depends on the cleanliness of the mobile phase and continued sealing capability of four check valves on each cycle (e.g. several times per minute).

Recent improvements include: A computer-designed camshaft is used to achieve maximum overlap of pump strokes, resulting in virtually undetectable pulsation or ripple and small-volume check valves are used to allow the pumps to function reliably at flow rates as low as 0.001 ml/min.

3.2. Injector

Sample introduction can be accomplished in various ways. The simplest method is to use an injection valve. In more sophisticated LC, automatic sampling devices are incorporated where sample introduction is done with the help of auto-samplers and microprocessors.

In liquid chromatography, liquid samples may be injected directly and solid samples need only to be dissolved in an appropriate solvent. The solvent need not to be the mobile phase, but frequently it is judiciously chosen to avoid detector interference, column/component interference or loss in efficiency. It is always best to remove particles from the sample by filtering, or centrifuging since continuous injection of particulate materials will eventually cause blockage of injection devices or columns.

Injectors should provide the possibility of injecting the liquid sample within the range of 0.1 to 100 ml of volume with high reproducibility and under high pressure (up to the 4000 psi). They should also produce minimum band broadening and minimize possible flow disturbances. The most useful and widely used sampling device for modern LC is the **microsampling injector valve**. With these sampling valves, samples can be introduced reproducibly into pressurized columns without significant interruption of flow, even at elevated temperatures.

With commercially available automatic sampling devices, large numbers of samples can be routinely analyzed by LC without operator intervention. Such equipment is popular for the analysis of routine samples (e.g., quality control of drugs), particularly when coupled with automatic data-handling systems. Automatic injectors are indispensable in unattended searching (e.g., overnight) for chromatographic parameters such as solvent selectivity, flow rate, and temperature optimization.

Most of the autosamplers have a piston metering syringe type pump to suck the preestablished sample volume into a line and then transfer it to the relatively large loop (~100 ml) in a standard six-port valve. The simplest autosamplers utilize the special vials with pressuarization caps. A special plunger with a needle, push the cap down to the vial and displace the sample through the needle into the valve loop. Most of the autosamplers are microprocessor controlled and can serve as a master controller for the whole instrument

3.3. Columns

The principle of ion exchange chromatography is that, charged molecules bind electrostatically to oppositely charged groups that have been bound covalently on the matrix. Ion exchange chromatography is a type of adsorption chromatography so that, charged molecules adsorb to ion exchangers reversibly so, the molecules can be bounded or eluted by changing the ionic environment. Ion exchangers can be used in column chromatography to separate molecules according to charge; actually other features of the molecule are usually important so that the chromatographic behavior is sensitive to the charge density, charge distribution, and the size of the molecule. An ion exchanger is usually a three-dimensional network or matrix that contains covalently liked charged groups. If a group is negatively charged, it will exchange positive ions and is a cation exchanger. An example of a group used in cation exchanger is the carboxy-methyl group. However, if a group is positively charged, it will exchange negative ions and is an anion exchanger. An example of a group used in anion exchanger is the di-ethyl-amino-ethyl group (DEAE). The matrix (stationary phase) can be made of various materials, commonly used materials are dextran, cellulose, and agarose.

The separation on an ion exchanger is usually accomplished in two stages: first, the substances to be separated are bound to the exchanger using conditions that give stable and tight binding; then the column is eluted with buffers of different pH, ionic strength or composition and the components of the buffer compete with the bound material for the binding sites. To choice whether the ion exchanger is to be anionic or cationic depend on the material to be separated. If the materials to be bound to the column have a single charge (i.e., either plus or minus), the choice is clear. However, many substances (e.g., proteins), carry both negative and positive charges and the net charge depends on the pH. In such cases, the primary factor is the stability of the substance at various pH values. Most proteins have a pH range of stability (i.e., in which they don't denature) in which they are either positively or negatively charged. So, if a protein is stable at pH value above the isoelectric point, an anion exchanger should be used; but if stable at values below the isoelectric point, a cation exchanger is required. Ion exchange columns vary widely in size, packing material and material of construction. Depending on its ultimate use and area of application, the column material may be stainless steel, titanium, glass or an inert plastic such as PEEK. The column can vary in diameter from about 2mm to 5 cm and in length from 3 cm to 50 cm depending on whether it is to be used for normal analytical purposes, microanalysis, high speed analyses or preparative work. The life of a column will depend largely on the type of samples it is used to separate but the conditions under which the separations are carried out will also place limits on it useful lifetime.

Guard column is placed anterior to the separating column. This serves as a protective factor that prolongs the life and usefulness of the separation column. They are dependable columns designed to filter or remove particles that clog the separation column and compounds and ions that could ultimately cause "baseline drift", decreased resolution, decreased sensitivity or create false peaks

3.4. Suppressor

The suppressor reduces the background conductivity of the chemicals used to elute samples from the ion-exchange column which improves the conductivity measurement of the ions being tested. IC suppressors are membrane-based devices which are designed to convert the ionic eluent to water as a means of enhancing the sensitivity. It can be used with universal detectors to act as a desalting device, thereby removing the interference resulting from the presence of ionic salts in the eluent. Suppressors are normally used with purely aqueous eluents, so there is a need to establish whether these suppressors can be used with the aqueous/ organic eluents needed to elute organic analytes which are retained on the stationary phase during their interaction. Eluents using ionic gradients and containing organic solvents can be suppressed satisfactorily using either chemical suppression with a micromembrane suppressor or electrolytic suppression using a self-regenerating suppressor. For utilization in industry, the electrolytic suppressor is usually more appropriate since it can employ water as the suppressor regenerant and is fully automated in terms of response to changing eluent conditions. Care needed to be taken with controlling the suppressor current in order to avoid damage to the suppressor and also the generation of ionic components from oxidation of the organic solvents (especially methanol) present in the eluent. Further potential problems, arising when using suppressors as de-salting devices with organic analytes, are the possibility of analytes loss in the suppressor as a result of adsorption or precipitation effects and dispersion of the analyte band in the suppressor.

Weakly acidic analytes are anionic in the presence of the high pH eluents used with anion-exchange IC, but become protonated in the suppressor and are therefore prone to hydrophobic adsorption or precipitation. Similarly, weakly basic analytes are separated as cations with low pH eluents but are deprotonated in the suppressor to form neutral species. The micro-membrane suppressor consists of layered ion-exchange membranes and fibrous chamber screens with the regenerant chamber screen modified to possess a high ion-exchange capacity which serves as a reservoir for regenerant ions. There is also a possibility of losses of analytes resulting from penetration of the analyte through the suppressor membrane into the regenerant chamber. Theoretically, anionic analytes are not able to penetrate the cation-exchange membranes of the anion suppressor due to the effects of Donnan exclusion.

Introduction of a suppression device between the column and the detector can be expected to cause some degree of peak broadening due to diffusional effects. The shape of the analyte band will also be influenced by hydrophobic adsorption effects, especially when the adsorption and desorption processes are slow. Examination of peak shapes and analyte losses can therefore provide important insight into the use of suppressors with organic analytes which are weakly acidic or weakly basic. It can be expected that peak area recovery rates after suppression are governed by a combination of hydrophobic interactions with the suppressor and permeation through the membranes with the balance between these mechanisms being determined by eluent composition, suppression conditions and analyte properties.

3.5. Detectors

Current LC detectors have a wide dynamic range normally allowing both analytical and preparative scale runs on the same instrument.

An ideal detector should have the following properties: low drift and noise level (particularly crucial in trace analysis), high sensitivity, fast response, wide linear dynamic range, low dead volume (minimal peak broadening), cell design which eliminates remixing of the separated bands, insensitivity to changes in type of the solvent, flow rate and temperature, operational simplicity and reliability. It should be non-destructive.

Electrical conductivity detector is commonly use. The sensor of the electrical conductivity detector is the simplest of all the detector sensors and consists of only two electrodes situated in a suitable flow cell. The sensor consists of two electrodes sealed into a glass flow cell. In the electric circuit, the two electrodes are arranged to be the impedance component in one arm of a Wheatstone bridge. When ions move into the sensor cell, the electrical impedance between the electrodes changes and the 'out of balance signal' from the bridge is fed to a suitable electronic circuit. The 'out of balance' signal is not inherently linearly related to the ion

http://dx.doi.org/10.5772/55652

concentration in the cell. Thus, the electronic circuit modifies the response of the detector to provide an output that is linearly related to the ion concentration.

The amplifier output is then either digitized, and the binary number sent to a computer for storage and processing, or the output is passed directly to a potentiometric recorder. This would result in a false change in impedance due to the generation of gases at the electrode surfaces. The frequency of the AC potential that is applied across the electrodes is normally about 10 kHz. In its simplest form, it can consist of short lengths of stainless steel tube insulated from each other by PTFE connecting sleeves.

Amperometric detection is a very sensitive technique. In principle, voltammetric detectors can be used for all compounds which have functional groups which are easily reduced or oxidized. Apart from a few cations (Fe^{3+}, Co^{2+}), it is chiefly anions such as cyanide, sulfide and nitrite which can be determined in the ion analysis sector. The most important applications lie however in the analysis of sugars by anion chromatography and in clinical analysis using a form of amperometric detection know as Pulsed Amperometric Detection (PAD).

Mass Spectrometry: Mass to charge ratio (m/z) allows specific compound ID determination. Several types of ionization techniques: electrospray, atmospheric pressure chemical ionization, electron impact. The detector usually contains low volume cell through which the mobile phase passes carrying the sample components.

Detector sensitivity is one of the most important properties of the detector. The problem is to distinguish between the actual component and artifact caused by the pressure fluctuation, bubble, compositional fluctuation, etc. If the peaks are fairly large, one has no problem in distinguishing them however, the smaller the peaks, the more important that the baseline be smooth, free of noise and drift. Baseline noise is the short time variation of the baseline from a straight line. Noise is normally measured "peak-to-peak": i.e., the distance from the top of one such small peak to the bottom of the next. Noise is the factor which limits detector sensitivity. In trace analysis, the operator must be able to distinguish between noise spikes and component peaks. For qualitative purposes, signal/noise ratio is limited by 3. For quantitative purposes, signal/noise ratio should be at least 10. This ensures correct quantification of the trace amounts with less than 2% variance. The baseline should deviate as little as possible from a horizontal line. It is usually measured for a specified time, e.g., 1/2 hour or one hour and called drift. Drift usually associated to the detector heat-up in the first hour after power-on.

Sensitivity can be associated with the slope of the calibration curve. It is also dependent on the standard deviation of the measurements. The higher the slope of your calibration curve the higher the sensitivity of your detector for that particular component, but high fluctuations of your measurements will decrease the sensitivity. The more selective the detection, the lower is signal/noise and the higher the sensitivity. The detector response is linear if the difference in response for two concentrations of a given compound is proportional to the difference in concentration of the two samples.

3.6. Data system

The main goal in using electronic data systems is to increase analysis accuracy and precision, while reducing operator attention. In routine analysis, where no automation (in terms of data management or process control) is needed, a pre-programmed computing integrator may be sufficient. For higher control levels, a more intelligent device is necessary, such as a data station or minicomputer.

4. Advanced applications of ion chromatography

Ion chromatography is basically a chromatographic method that has become a routine analytical method. It is regarded as a versatile analytical technique for separating and quantifying ions. The concept of IC was successively widened with advancements of the rapid development in separation, column stationary phase, great variety of detectors, data analysis and hyphenated techniques. Moreover, it could include other separation methods (e.g., ion interaction and ion exclusion) for simultaneous separation of analyte components. IC analysis has matured to a well-established rugged, sensitive and reliable analysis technique for a wide variety of chemical compounds present in various matrices. On this manner, many papers have been published during the last few years dealing with new modalities in sample pretreatment, separation, detection, etc., for improving samples analysis. The following section deals with the recent development in instrumentations and applications to fit the desired fields of applications.

4.1. Qualitative and quantitative analysis of cations and anions

The demand for the determination of ionic species in various water samples is growing rapidly along with increasing environmental problems and it is clearly important to develop an appropriate analytical method for their determination. IC represents one of the most efficient methods that provide accurate and rapid determination of ionic species in water samples. Basically, anions and cations can be independently separated. Recent advances in ion chromatography (IC) make it a superior analytical method; it has been expanded for the simultaneous determination of inorganic anions and cations. Column switching has become a capable technique for the simultaneous determination of inorganic anions and cations in a single chromatographic run. Amin et al. [10] demonstrated a convenient and applicable method for various natural fresh water samples analysis (Figure 5). They proposed an ion chromatography (IC) method for the determination of seven common inorganic anions (F^-, $H_2PO_4^-$, NO_2^-, Cl^-, Br^-, NO_3^-, and SO_4^{2-}) and/or five common inorganic cations (Na^+, NH_{4+}, K+, Mg^{2+}, and Ca^{2+}) using a single pump, a single eluent and a single detector. The system used cation-exchange and anion-exchange columns connected in series via a single 10-port switching valve. The 10-port valve was switched for the separation of either cations or anions in a single chromatographic run. Using a specific eluent, 1.0 mM trimellitic acid (pH 2.94), seven anions and the five cations could be separated on the anion-exchange column and the cation-exchange column, respectively. The elution order was found to be $F^- < H_2PO_4^- < NO_2^- < Cl^- < Br^- < NO_3^-$

< SO_4^{2-} for the anions and Na^+ < NH_4^+ < K^+ < Mg^{2+} <Ca^{2+} for the cations. Complete separation of the above anions or cations was demonstrated within 35 min each. Detection limits calculated were 0.05–0.58 ppm for the anions and 0.05–0.38 ppm for the cations, whereas repeatability values were below 2.26, 2.76, and 2.90% for peak height, peak area and retention time, respectively.

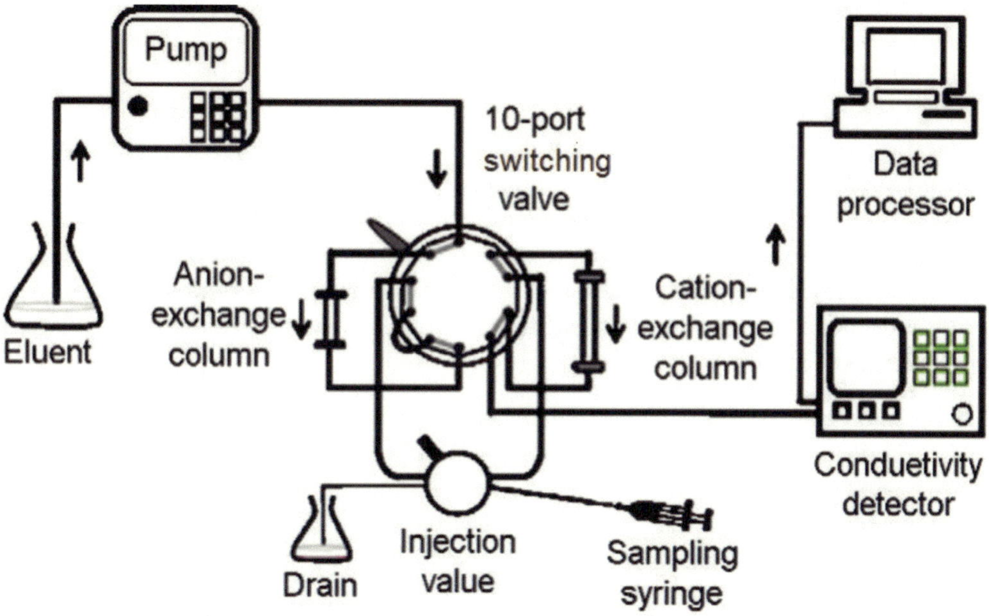

Figure 5. Schematic diagram of the instruments used for simultaneous separation of anions and cations [10].

4.2. Qualitative and quantitative analysis of halides

4.2.1. Bromate

Bromate has been classified as a human carcinogen by both the IARC (International Agency for the Research on Cancer) and the USEPA (United States Environmental Protection Agency) and is known to be toxic to fish and other aquatic life [11, 12]. Bromate could be produced in aquatic systems upon the oxidation of aqueous bromide. Controlled ozonation has been considered as an effective disinfectant tool in aquatic systems [13] but when sea water is subjected to ozonation, oxy-bromide ozonation by-products (OBP) are produced and these are important both in terms of their disinfection ability and also in relation to their potential toxicity. When seawater is oxidized, aqueous bromide (Br-) is initially converted to hypobromite (OBr^-) which can then either be reduced back to bromide or oxidized further to bromate (BrO3-) which is known to be toxic to fish and other aquatic life and classified as a human carcinogen. There has been thus a considerable interest in bromate analysis so that trace analysis of bromate in water has received considerable attention in recent years.

Zakaria et al. [12] used a multi-dimensional matrix-elimination ion chromatography approach, two-dimensional and three-dimensional configurations as described in Figure 6, for the determination of bromate in seawater samples. The designed configurations were used effectively to eliminate the interference caused by the high concentration of ubiquitous ions present in seawater such as chloride and sulfate. A two-dimensional approach utilizing a high capacity second dimension separation, comprising two columns connected in series, was applied successfully and permitted the determination of bromate in undiluted seawater samples injected directly onto the ion chromatography system. A three-dimensional method utilizing two 10-port switching valves (Figure 6b) to allow sharing of the second suppressor and detector between the second and third dimension separations showed better resolution and detection for bromate and reduced the limit of detection to 60 μg/L for spiked seawater samples. Experimentally, the analyzed ozonated seawater samples exhibited a non-linear increase in bromate level on increasing ozonation time. A bromate concentration in excess of 1770 μg/L was observed following ozonation of the seawater sample for 120 min. The developed method provides the elimination of high concentration of interfering species, such as chloride and sulfate, by taking specific fractions from each separation column and re-injecting onto a subsequent column.

Using this approach, the limit of detection for bromate was 1050 μg/L using a 500 μL injection loop. Good linearity was obtained for bromate with correlation coefficients for the calibration curves of 0.9981 and 0.9996 based on peak height and area, respectively. The limit of detection achieved was more than sufficient to determine levels of bromate known to be toxic to aquatic species of interest in aquaculture applications. The developed method is therefore applicable to aquaculture, especially where water is recycled and repeatedly ozonated, leading to the probability of accumulation of bromate. Furthermore the described method is generally applicable to other high ionic strength samples, although re-optimization of cutting times would be required. The system is also potentially applicable for the analysis of other low concentration ionic species, including other oxyhalides such as chlorate.

4.2.2. Iodide and iodate

One of the problems of iodide estimation by conductivity detection is the expected interference from other ions and poor sensitivity of detection which rendered its estimation in complex samples difficult to apply. On the other hand, several methods have been developed for the estimation of iodate ion in water, however, one drawback of these methods is that it can give false estimation of iodate with oxidizing agents such as bleaching powder, which too can generate iodine from the reaction with I^-. It is therefore necessary to devise a sensitive and selective precise test for the separation and detection of iodate species in different samples matrices. Kumar et al. [14] applied successfully an ion chromatographic method with conductivity detection for iodate estimation in common salt after sample pretreatment with on-guard silver cartridge for the removal of the large excess of chloride ion. Unfortunately, fresh Ag cartridge is required for each sample which would render the method expensive for routine use.

http://dx.doi.org/10.5772/55652

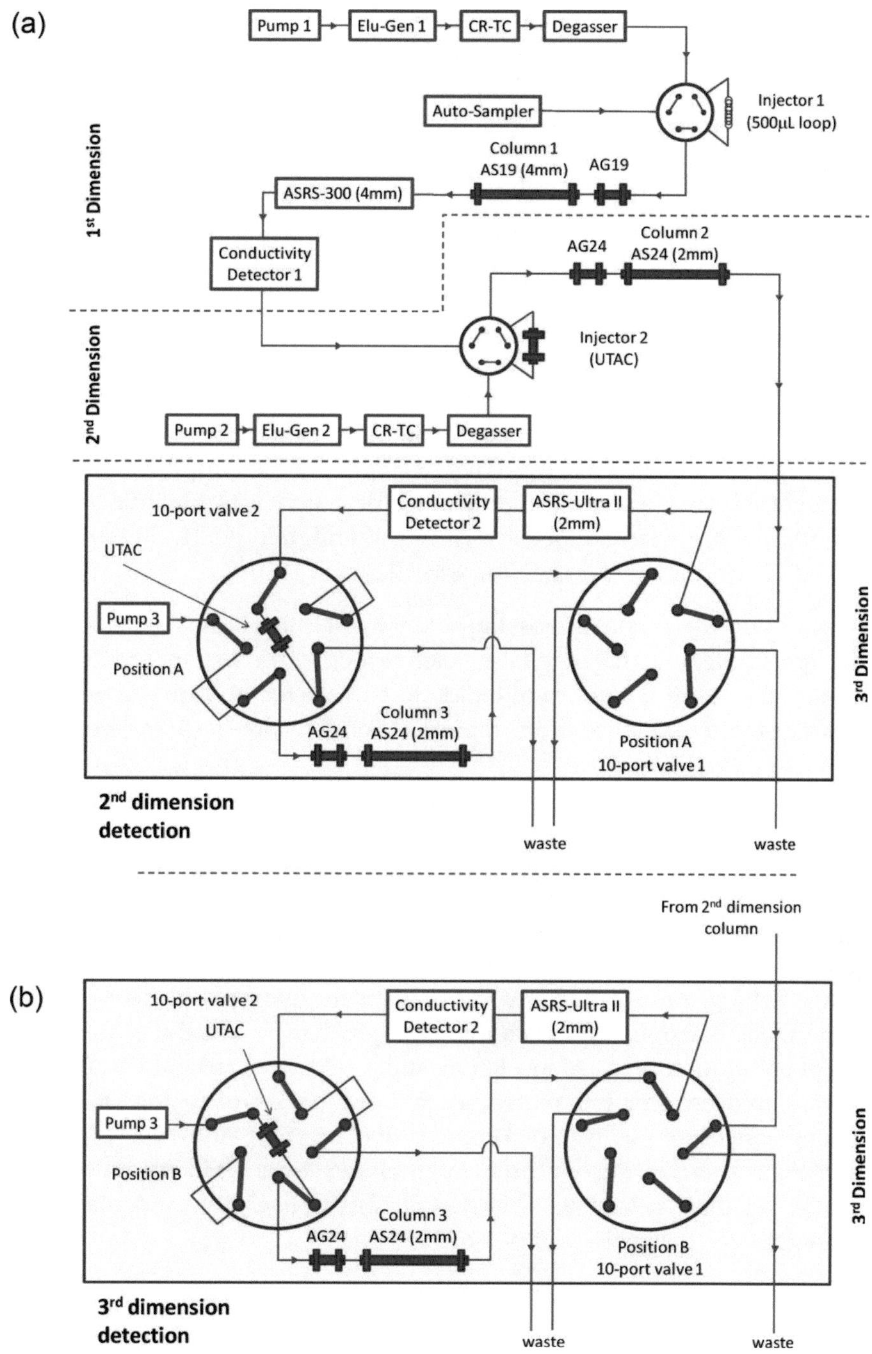

Figure 6. Schematic diagram of instrumentation used to perform the multi-dimensional IC. (a) 10-Port valve positions for detection of 2nd dimension separation (i.e. effluent from column 2 diverted through conductivity detector 2). (b) 10-Port valve positions for injection of 2nd dimension cut fractions onto 3rd dimension column with subsequent detection using conductivity detector 2 [12].

Ion chromatography employing anion-exchange column with amperometric detection is demonstrated to be well suited for quantitative estimation of iodide and iodate in iodised salt [15].The success of the technique, which dispenses with the need for pre-treatment for chloride removal, hinges on the excellent resolution achieved and the high selectivity and sensitivity of detection of iodide by amperometry. The system consisted of a gradient pump with vacuum degas option, an electrochemical detector, liquid chromatography module, eluant organiser and rheodyne injection loop and PVDF (polyvinylidene fluoride) filters with pore size of 0.45 μm. The flow-through detection cell is made of a 1.0-mm diameter silver working electrode and a pH-Ag | AgCl combination reference electrode. The titanium body of the cell served as the counter electrode. Separations were accomplished on a 250 mm × 4 mm i.d. column coupled with a 50 mm × 4 mm i.d. guard column. Such a column contains a hydrophilic, anion-exchange resin that is well suited to the chromatography of the relatively hydrophobic iodide anion. Elution was carried out under isocratic condition using HNO_3 (50 mM) at a flow rate of 1.5 mL/min. The injection loop volume was fixed at 50 μL and the sample run time was 10 min. Ion chromatographic analysis with conductivity detection was undertaken on the same column using 22 mM NaOH as eluant and flow rate of 1 mL/minute. The injection loop volume was fixed at 10 μL and the sample run time was 10 min.

This technique is easy to use and its most important merit is that it can readily indicate absence of iodate in case adulterants that give false positive iodometric test are used in its place. The method also enables trace quantities of iodide to be detected even in the presence of large excess of chloride ion. Interferences from impurities normally present in salt were insignificant.

4.2.3. Perchlorate

In chromatographic analysis, the highly retained species present a challenge for ion chromatographic analysis due to peak broadening which leads to low resolution between analytes of interest and to relatively poor detection limits. This problem is often more acute with monovalent anions than with monovalent cations because common anions are often large, and the greater radius to charge ratio facilitates partitioning to the hydrophobic stationary phase. The introduction of macrocycle-based ion chromatography has provided useful new techniques for analysis of both cations and anions. For example, capacity gradient ion chromatography [16] is beneficial in decreasing retention times and thus peak broadening for highly retained anions, making possible the analysis of a broad host of anions. Lamb et al. [17] focused on the introduction of macrocycles into ion chromatographic systems for increased versatility in the separation of both cations and anions. They described extensively the use of macrocycles based ion chromatography in the analysis of perchlorate ion.

As more information on the extent of the contamination and the dangerous effects of perchlorate consumption has become available, much concern has arisen over perchlorate contamination in public water systems. Furthermore, the US Environmental Protection Agency (USEPA) has periodically reduced the acceptable limit for safe consumption. Currently, the limit stands at 0.7 μg/kg/day, which corresponds to 24.5 μg/L for a 70 kg human drinking 2 L of water per day. The method described by Lamb et al. [17] provides effective perchlorate determinations (shown in Figure 7) using standard conductimetric detection by combining an

18-crown-6-based mobile phase with an underivatized reversed-phase mobile phase ion chromatography (MPIC) column. One unique feature of this method is the flexibility in column capacity that is achieved through simple variations in eluent concentrations of 18-crown-6 and KOH, facilitating the separation of target analyte anions; perchlorate. Using a standard anion exchange column as concentrator makes possible the determination of perchlorate as low as 0.2 μg/L in low ionic strength matrices. Determination of perchlorate at the sub-ug/L level in pure water and in hardwater samples with high background ion concentrations can be achieved this way. However, like other IC techniques, this method is challenged to achieve analyses at the μg/L level in the demanding high ionic strength matrix described by the United States Environmental Protection Agency (USEPA) (1000 mg/L chloride, sulfate and carbonate) [17]. This challenge was approached by use of the Cryptand C1 concentrator column to effectively preconcentrate perchlorate while reducing background ion concentrations in the high ionic strength matrix. The method makes possible the determination of perchlorate at the 5 μg/L level in the highest ionic strength matrix described by the EPA. In short, this method provides an alternative method for the analysis of perchlorate at concentration levels as low as 5 μg/L in high background samples and to well below 1 μg/L in pure water and low salt samples.

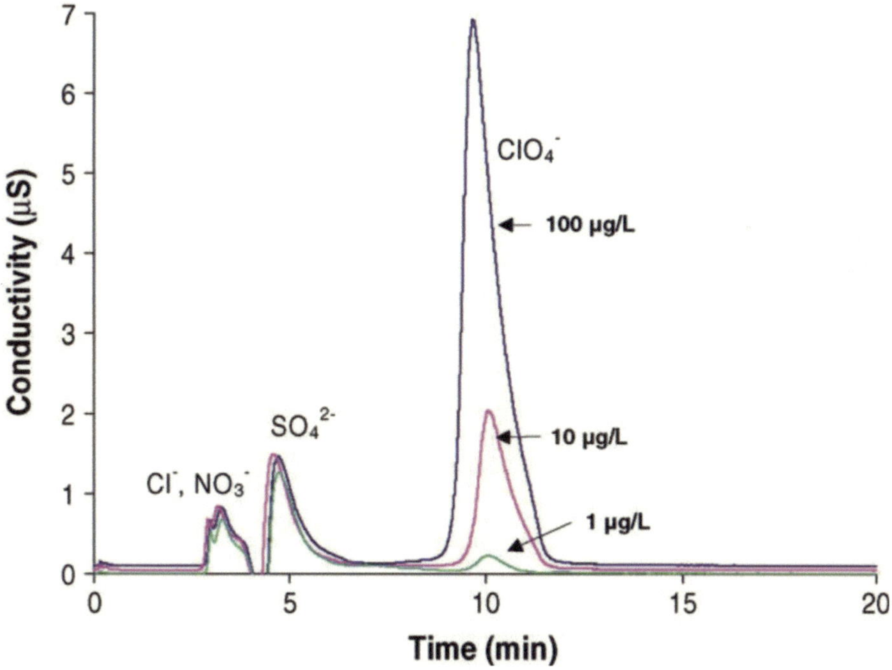

Figure 7. Optimal system configuration using AG4 guard column as the concentrator column. Five milliliters of Milli-Q water spiked with $ClO4^-$ was loaded onto the concentrator column at varying concentrations of perchlorate. Eluent: 0.5M 18-crown-6 and 5mM KOH. Injection: 5mL loaded onto concentrator column, flow rate: 1.0 mL/min, temperature: 20 °C [17].

4.3. Trace and ultra-trace analysis

There has been considerable interest in the determination of ions at trace levels as, for example, in applications need high-purity water as in semiconductor processing and the determination of trace anions in amine treated waters. For this investigation, we will define "trace" as determinations at or below 1 μg/l (ppb) levels. The Semiconductor Equipment and Materials International (SEMI) recommended the use of IC for tracking trace ionic contaminants from 0.025 to 0.5 μg/l [18]. In addition, the Electric Power Research Institute (EPRI) has established IC as the analytical technique for determining of trace level concentrations of sodium, chloride and sulfate down to 0.25 μg/1 in power plant water [19].

To determine ions at mid μg/l to mg/l (ppb to ppm) levels with IC, a sample size of 10 to 50/μl is sufficient. To determine ions at lower levels, then a preconcentration or trace enrichment technique has typically to be utilized [20]. With this method, the analytes of interest are preconcentrated on another column in order to "strip" ions from a measured sample volume. This process concentrates the desired species resulting in lower detection limits. However, preconcentration has several disadvantages, compared with a direct method, additional hardware is required. A concentrator column is used to preconcentrate the ions of interest, a sample pump is needed for loading sample, an additional valve is often required for switching the concentrator column in and out-of line with the analytical column and extra time is required for the preconcentration step. It was of interest to explore the development of a high-volume direct-injection IC method that would facilitate trace ion determinations without a separate preconcentration step. This would represent a significantly simpler and more reliable means of trace analysis.

Kaiser et al. [21] described the evaluation of on-column preconcentration for enhancing sensitivity and enabling trace ion determination in high-purity water. They developed a high-volume direct-injection method for trace level determinations (low to sub μg/l) of anions and cations by ion chromatography as shown in Figure 8. The chromatographic signal was enhanced by increasing the sample volume up to 1300/μl with no significant loss in peak efficiency. Total analysis times were less than 30 min and the method detection limits for most ions ranged from 10 to 400 ng/l (ppt). The methods described exhibit increased sensitivity and greater reliability than methods using conventional preconcentration. Lower detection limits were achieved by increasing sample size with no significant loss neither in peak efficiency nor in peak resolution. Trace levels (low to sub μg/l) were determined without the added complexity of a concentrator column or loading pump and valve.

4.4. Heavy metals

Błazewicz et al. [22] investigated the basic validation parameters of determining the transition metal ions using ion chromatography. Moreover, they described the use of IC method together with the digestion conditions for the determination of heavy metals in different solid matrices. They designed an ion chromatography preceded by microwave-assisted acidic digestion of tissues samples in appropriate conditions for the determination of Co^{2+}, Cu^{2+}, Fe^{3+}, Mn^{2+}, and Zn^{2+} in human tissues (nodular goitre and healthy human thyroids). Microwave mineralization in a closed system, where the contamination problems are significantly reduced, is recom-

http://dx.doi.org/10.5772/55652

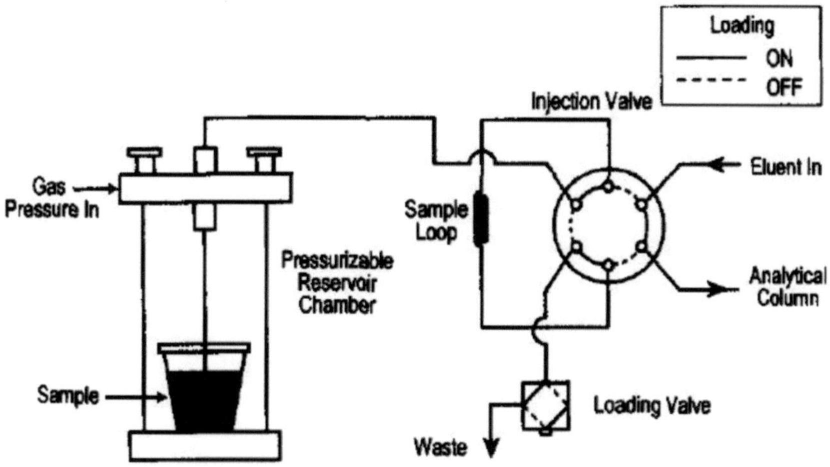

Figure 8. IC system configuration for direct-injection sample loading [21].

mended for such samples. The chromatogram of heavy metals separation is represented in

Figure 9.

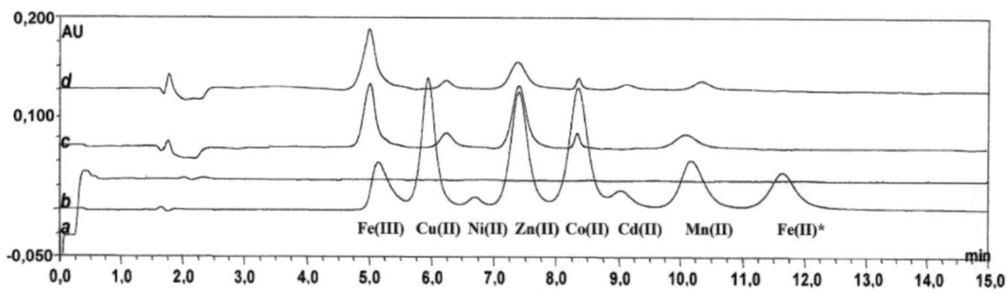

Figure 9. Chromatograms of blank sample (a), standard mixture (b), sample of thyroid from the control group (c), and sample of thyroid of a patient with diagnosed nodular goitre (d) [22].

Metal ion	LOD (LOQ)* (μg/mL)	Measured value (μg/g) ± SD	Certified value (μg/g) ± SD
Cd	0.022 (0.073)	3.75 ± 0.45	4.15 ± 0.38
Co	0.026 (0.087)	0.66 ± 0.15	0.57 ± 0.11
Cu	0.048 (0.16)	65.1 ± 5.9	66.3 ± 4.3
Fe	0.009 (0.03)	586 ± 17.2	539 ± 15
Mn	0.006 (0.020)	11.9 ± 2.1	12.3 ± 1.5
Ni	0.006 (0.020)	2.14 ± 0.59	2.25 ± 0.44
Zn	0.056 (0.19)	857.2 ± 45	830 ± 57

* LOD—limit of detection; LOQ—limit of quantitation; LOQ= 10/3 LOD.

Table 1. Comparison of metal ions concentrations measured by IC and certified values [22].

An evaluation of the obtained data indicated that the mean values found for iron, copper, and zinc are within the values presented in literature. The main assets of the presented method lie in its simplicity and the practicality of determining analytes from samples of various origins. Suitability of the developed IC method was supported by validation results as shown in Table 1. Generally, very good results of precision (RSD below 5%) and recoveries (above 90%) were evaluated.

On the other hand, in the framework of long-term management and recycling of nuclear wastes, the transmutation process has been identified as a promising option to decrease the radiotoxicity of radionuclides. A sample containing around 5 mg of ^{109}Ag metal powder is one of the fission product transmutation targets which were irradiated. This sample was initially enriched in ^{109}Ag (>99%). After irradiation, the theoretical evolution scheme predicts respectively the formation of 366 μg of cadmium and 1 μg of palladium compared to 4636 μg of silver. Determination of cadmium isotopic compositions is of prime interest to validate neutron calculation codes and to obtain the integral capture cross section of ^{109}Ag. Isobaric interferences occur at mass 108 between cadmium and silver, and at mass 110 between cadmium, silver and palladium. The mass resolution required to overcome ^{110}Cd/^{110}Ag/^{110}Pd interference is about 100,000 which is beyond actual possibilities of mass spectrometers. Thus, a chemical separation step must be completed to isolate cadmium in a purified fraction before offline isotopic measurements. In the case of radioactive materials, the chemical separations performed with gravity flow on ion exchange resins induce drawbacks for analysts, such as increased handling time on samples. Moreover, in most proposed procedures, cadmium is generally eluted after silver, which lowers the separation factor between silver and cadmium and decreases decontamination factors because of silver peak tailing in cadmium fraction. A powerful way to reduce analysis time and to improve selectivity is high performance ion chromatography. However, no detector classically associated with a HPIC system can measure Ag, Cd and Pd with high specificity and sensitivity. Ion chromatography-inductively coupled plasma mass spectrometry (IC-ICPMS) can tackle those specifications: it can be used to detect trace elements at the exit of the chromatographic column. Because of the fast mass scanning ability of the quadrupole in peak jumping mode, this kind of spectrometer enables an easy handling of transient signals associated with high sensitivity. The separation procedure was achieved with a carboxylate-functionalized cation exchange CS12 column using 0.5 M HNO_3 as eluent giving satisfactory results in terms of peak resolution and decontamination factors for example. The developed method demonstrates the possibility to obtain rapidly purified cadmium fractions which can be directly analyzed by multi collection inductively coupled plasma mass spectrometry (MC ICPMS). After the optimization of chromatographic conditions, the method was applied to the separation of non-radioactive solutions simulating the composition of the irradiated sample. This hyphenated technique minimizes sample pretreatment and shortens analysis time, which is of prime interest for nuclear applications. Moreover, the developed method displays a strong potential not only for nuclear issues but also for geological and cosmochemical applications where high accuracy and precision isotopic analyses are also needed [23].

4.5. Inorganic compounds

Hydrogen cyanide (HCN) is one of the major ciliatoxic components when tobacco products such as cigarettes are combusted and is thus classed to the "Hoffmann analytes" It is formed in cigarette smoke in the burning zone mainly from pyrolysis of various nitrogenous compounds such as protein and nitrate in tobacco at oxygen-deficient conditions. The quantitative determination of HCN in cigarette smoke is integral to proper assessment due to its potential impact on public health. However, there are many challenges in accurately determining its amount in cigarette smoke; these include the need for developing an efficient and rapid smoke collecting method and the technique to analyze it in complex smoke mixture. Recently, extensive efforts have been done on determining cyanide by IC through the development and application of electrochemical detection (especially pulsed amperometric detection) which endow this kind of method with a high selectivity and improved accuracy, which eventually enable them to be widely applicable to the ion chromatography. Zhang et al. [24] focused on applying ion chromatography in the determination of hydrogen cyanide in cigarette main stream smoke. Whole cigarette mainstream smoke was totally trapped by Cambridge filters, which are treated with sodium hydroxide/ ethanol solution as shown in Figure 10.

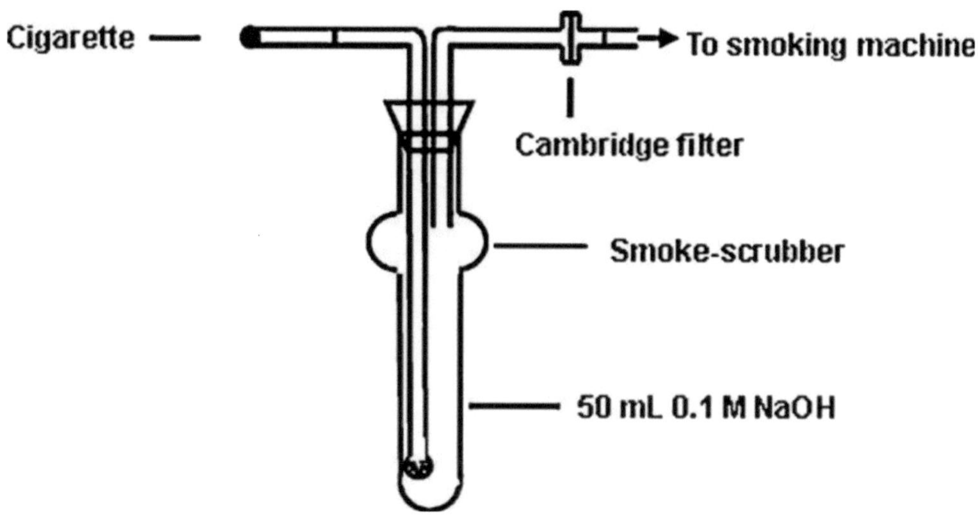

Figure 10. Schematic diagram of solution absorption method [24].

The chromatographic analysis (Figure 11) has been achieved by developing an ion chromatography integrated with pulsed amperometric detection (PAD) and optimizing some factors include sample treatment, matrix interference, composition of eluents and so on. The method possesses the advantage of fast analysis time over the widespread used solution absorption method. The possible co-existing interferents are evaluated under the optimized detection conditions and excellent recoveries of cyanide. The optimization of composition of eluents and evaluation of possible interferents make this method selective and reliable so that the cyanide content of absorption solution can be directly determined by the optimized IC-PAD method without any pretreatments. The linear range is 0.0147–2.45μg/mL with R^2 value of 0.9997. The limit of the detection is 3μg/L for a 25μL injection loop. The overall relative standard deviation of the method is less than 5.20% and the recovery range from 94.3% to 101.0%. This developed method proves to be advantageous, due to expanded detection range with greater accuracy and is thus highly anticipated to find wide applications in cigarette smoke analysis.

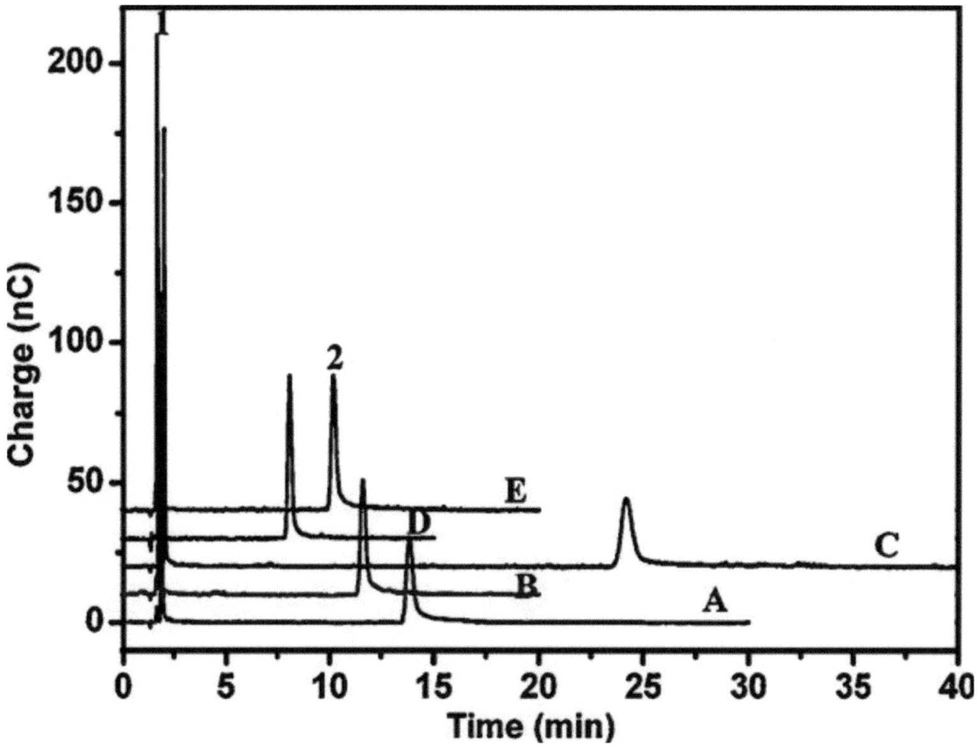

Figure 11. Typical chromatograms obtained under the following eluent composition: A: 0.2M NaOH, 0.2M NaAC; B: 0.4M NaOH, 0.2M NaAC; C: 0.6M NaOH; D: 0.6M NaOH, 0.3M NaAC; E: 0.6M NaOH, 0.2M NaAC. Flow rate: 1.0 mL/min, injection volume: 25 μL, column temperature 30°C. Peak 1: unidentified, peak 2: cyanide [24].

4.6. Organic compounds

4.6.1. Hippuric acid

IC can also be used in detection of some acids. Zhao et al. [25] proposed a simple and eco-friendly ion chromatographic method for the determination of Hippuric acid (HA) in human urine (see Figure 12). Hippuric acid is a kind of metabolite of toluene in human body, therefore, HA is a physiological component of human urine if toluene was inhaled. The content of HA in human urine actually is confirmed as a diagnostic marker of exposure to toluene [26]. It has been reported that exposure to high concentrations of volatile organic compounds such as toluene lead to a series of diseases such as acute and chronic respiratory effects, functional alterations of the central nervous system, mucous and dermal irritations, and chromosome aberrations.. In order to diagnose patients who are suffering from a series of diseases caused by elevated HA levels, the determination of HA in human urine is necessary. Comparing with other chromatographic methods such as GC and HPLC, the proposed IC method used eco-friendly mobile phase (not containing organic solvent), and avoided complicated sample pretreatment. The separation was carried out on an anion exchange column with 2.0 mmol/L $NaHCO_3$ as mobile phase at the flow-rate 0.7mL/min. A suppressed conductivity detector was used and the detection limit was 1.0 μg/L (S/N = 3) for hippuric acid. The analysis time for one run was 30 min under the optimized IC condition. The recovery of hippuric acid was 93.2–98.0% while the relative standard deviation (RSD) was 1.4–2.3% by seven measurements. Furthermore the results shown that the proposed method has the advantages of easy operation, high sensitivity and accuracy. This method is suitable for routine clinical analysis of HA.

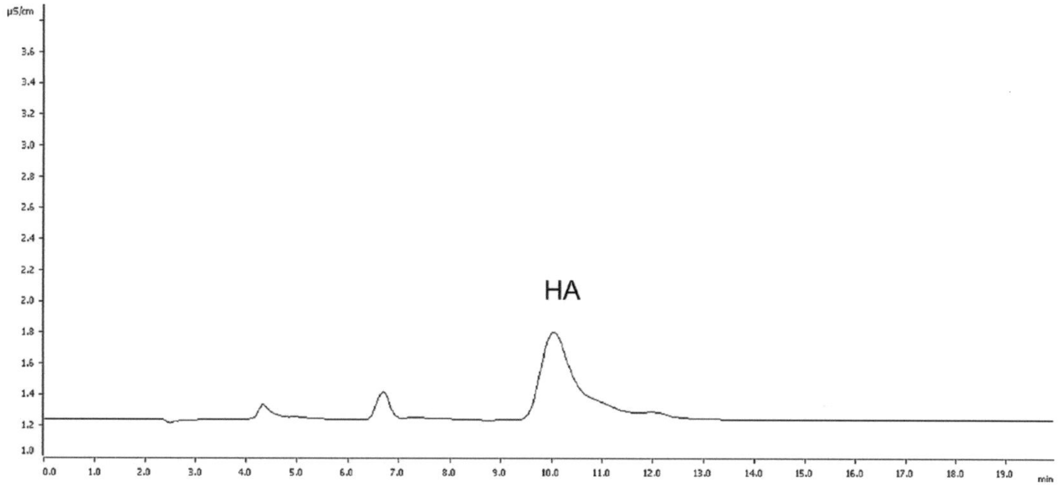

Figure 12. Chromatogram of a standard solution of HA (10mg/L) [25].

4.6.2. Amines and its derivatives

Erupe et al. [27] developed an ion chromatography method with non-suppressed conductivity detection for the simultaneous determination of methylamines (methylamine, dimethylamine, trimethylamine) and trimethylamine-N-oxide (TMAO) in particulate matter air samples. The method can be used to detect, quantify and determine whether TMAO and methylamines are quantitatively significant components of organic nitrogen aerosol in the atmosphere. This was done using aerosol collected from smog chamber reactions of trimethylamine with ozone and/or nitrogen oxide. The method was tested using a solution of laboratory-generated aerosol containing a mixture of the analytes. The analytes were well separated by means of cation-exchange chromatography using a 3 mM nitric acid / 3.5% acetonitrile (v/v) eluent solution and a Metrosep C 2 250 (250mm×4mm i.d.) separation column. The composition of the mobile phase was optimized and efficient separations between the analytes were achieved (Figure 13 and 14). Detection limits of methylamine, dimethylamine, trimethylamine, and trimethylamine-N-oxide were 43, 46, 76 and 72 µg/L, respectively. The method described is simple and has low detection limits suitable for analysis of aerosols generated in smog chamber experiments and in ambient air where the concentration of these species is expected to be high.

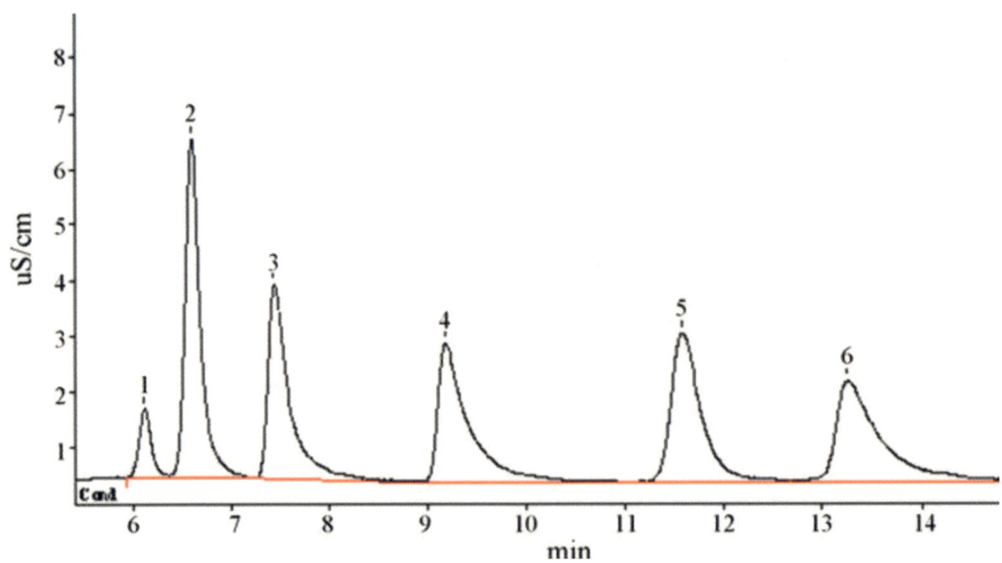

Figure 13. Separation of methylamines and methylamine-N-oxides from standard solutions. Analytes: 1-sodium, 2-ammonium, 3-methylamine (195 µg/L), 4-dimethylamine (390 µg/L), 5-trimethylamine-N-oxide (465 µg/L), and 6-trimethylamine (615 µg/L) [27].

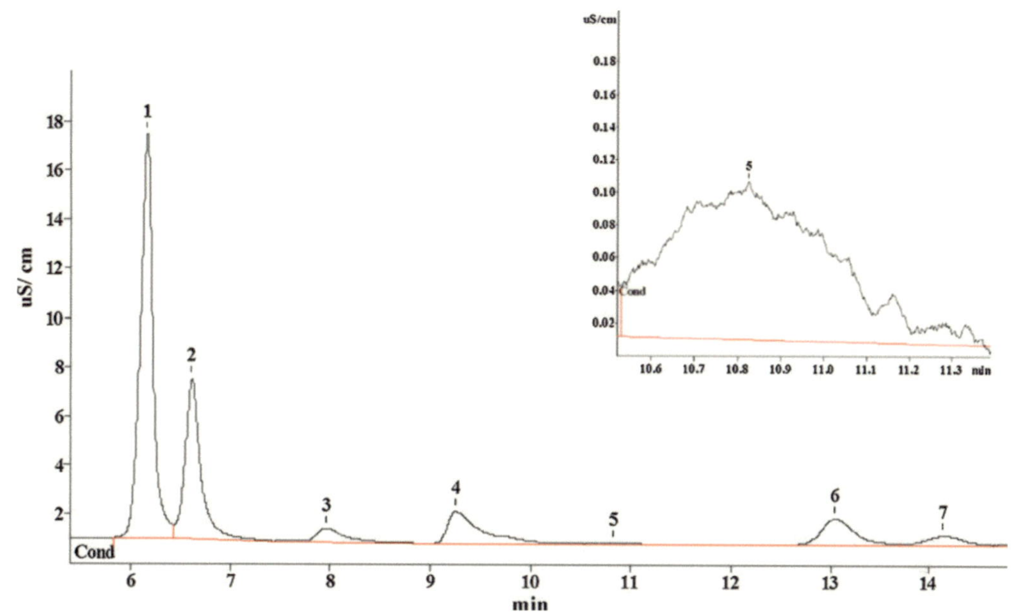

Figure 14. Chromatogram of smog chamber filter analysis from reaction of trimethylamine with ozone. Analytes: 1-sodium, 2-ammonium, 3-potassium, 4-dimethylamine (1.72 µg/m^3), 5-trimethylamine-N-oxide (0.25 µg/m^3), 6-magnesium, and 7- trimethylamine (0.57 µg/m^3). The inset is a magnification of the trimethylamine-N-oxide peak (5) from the chromatogram [27].

4.6.3. Phenolic compounds

Phenolic compounds have attracted great concern in recent years due to their high toxicity and bio-recalcitrant effect in the ecosystem water cycling process. Numerous techniques have been studied and developed to determine phenols. However, most of these detection techniques focuses on high performance liquid chromatography (HPLC) equipped with various kinds of detectors such as UV, electrochemical, fluorescence, and mass spectroscopy [28,29]. Among these detection techniques, fluorescence detector is a better choice in terms of selectivity and sensitivity. HPLC combined with fluorescence detector (HPLC/FD) have been used in numerous applications in trace analysis. However, some phenols have weak fluorescent property and post-column derivatization is often required to convert these compounds into strong fluorescent substances that can then be efficiently detected by the fluorescence detector [30]. Using on line electrochemical derivatization, Karst et al. [30] presented a method to determine mono-substituted phenols via HPLC equipped with fluorescence detector (HPLC/ED/FD). This method addressed the problems on phenols that could not be detected via fluorescence detector. However, the separation was performed by common silica-based C18 separation column. Unfortunately, the silica column works well only in the pH range of 2–8 (pH < 3), whereas the optimum pH for producing the fluorescence of oxidized phenols is basic (pH ~10). Obviously, the separation condition could not match well with that of downstream detection. Therefore, buffer solution of NH_3/NH_4Cl at pH 9.5 had to be added to the effluent from the column to perform the electrochemical conversion to enhance the fluorescence signal. Polymer-based stationary phases (e.g. divinylbenzene/ ethylvinylbenzene, DVB/EVB) in IC dominate most of the applications due to their wide pH tolerance (0–14). Since the polymer-based column can work well in alkaline solution (e.g., pH ~10). The choice of alkaline eluent matching with the downstream fluorescence detection will not be a barrier if the phenols could be well separated by IC. Based on these considerations, a method to determine phenols, where their separation is performed using IC combined with online post-column, electrochemical derivatization and fluorescence detection (IC/ED/FD), has been developed [31] Six model phenols including 4-methylphenol (pMP), 2,4-dimethylphenol (DMP), 4-tert-butylphenol (TBP), 4-hydroxylphenolacetic acid (pHPA), 4-acetamidophenol (pAAP), and phenol (P) were well separated on an anion-exchange column under ion exchange mode using NaOH with small amount of acetonitrile added as eluent (as shown in Figure 15). The separation of phenols was carried out in the anion exchange column with basic eluent and the electro-oxidation of phenols is performed using a laboratory-made electrolytic cell (EC) consisting of porous titanium electrode and cation exchange membrane (CEM) which allows the oxidation products that are strongly fluorescent to be detected by the fluorescence detector. NaOH eluent used in the present method matches well with the maximal fluorescence intensity obtained at alkaline condition for oxidized phenols, thus the addition of specific buffer solution after oxidation could be eliminated. This method leads to a simplified procedure and eliminates the use of additional setup and greatly simplifies the operating procedures. The proposed method was sensitive to the limits of detection in the range of 0.4 μg/L and 3.8 μg/L and the limits of quantification between 1.2 μg/L and 13 μg/L due to the strong electro-oxidation capacity of porous titanium electrode, as well as the implementation of time-programmed potential over EC. The linear ranges were $2.0–1.0 \times 10^4$ μg/L for pAAP and DMP, and $10–1.0 \times 10^4$ μg/L for P, pMP, pHPA, and TBP, respectively. The relative standard deviations range from 0.9% to 4.8%. The utilization of the method was demonstrated by the analysis of real samples.

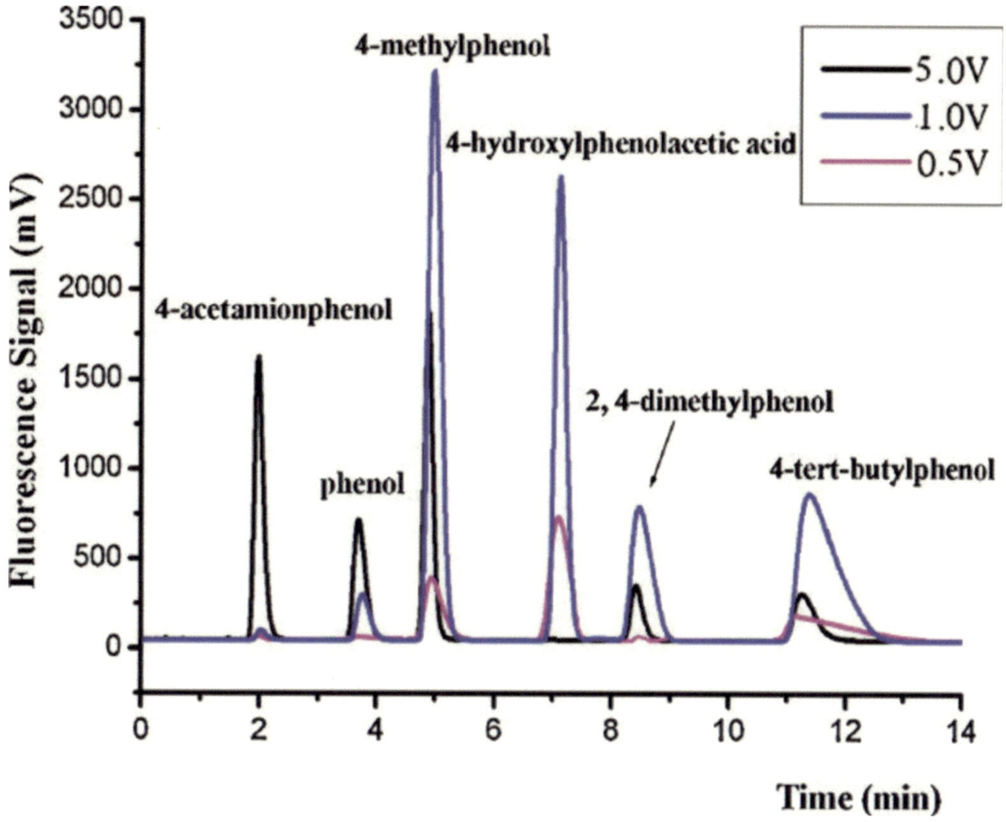

Figure 15. Chromatograms of phenolic compounds (~10 mg/L) at different potential [31].

5. Conclusion

This chapter deals with ion exchange chromatography, IC, as a subset of liquid chromatography. Due to the continuous growth, chromatography became one of the most widely used methods in different branches of science encompassing chemistry, physical chemistry, chemical engineering, biochemistry and cutting through different fields of analytical proposes.

Discovery and historical background on IC were mentioned. Steps of ion chromatography process were intensively discussed in addition to instrumental components of typical IC instrument including: pump, injector, column, suppressor, detector and recorder or data system.

The chapter emphasizes the superior analytical power of ion chromatography so that it can be used for qualitative and quantitative analysis of common cations, anions and halides in their different forms and matrices in trace and ultra-trace concentrations. Heavy metals separation and detection was also mentioned as well as hydrogen cyanide as an example of inorganic compounds. As examples of organic acid separation and detection using ion chromatography, the analysis of hippuric acid, amines and its derivatives and phenolic compounds were mentioned.

Author details

Yasser M. Moustafa* and Rania E. Morsi

*Address all correspondence to: ymoustafa12@yahoo.com

Egyptian Petroleum Research Institute, EPRI, Cairo, Egypt

References

[1] Handbook of Ion Chromatography Third, Completely Revised and Enlarged Edition, Joachim Weiss (2004).

[2] Tswett M. S., Khromofilly v Rastitel'nom i Zhivotnom Mire (Chromophylls in the Plant and Animal World). Karbasnikov Publishers, Warsaw (1910).

[3] Gjerde D. T. & Fritz J. S. Ion Chromatography (2000). Weinheim: Wiley-VCH.

[4] Basic principles of Ion chromatography: IC Module Navigation, The University of MACHIAS at MACHIAS.

[5] Theory of Ion Chromatography, Mterohm, UK, LTD.

[6] Ion chromatography, Hamish Small, online book, Springer, Amazon.com

[7] Ion exchange chromatography: Principles and Methods, Amersham Bioscience.

[8] Chromatography: a laboratory handbook of chromatographic and electrophoretic techniques Heftman, E. (Ed.), Van Noostrand Rheinhold Co., New York (1975).

[9] Giddings J. C. & Keller R. A. Dynamics of chromatography, Part 1, Principles and theory (1965). Marcel Dekker Inc., New York.

[10] Amin M., Lim L. W & Takeuchi T. Determination of common inorganic anions and cations by non-suppressed ion chromatography with column switching. Journal of Chromatography A 2008; 1182 (2) 169-175.

[11] Grguric G., Trefry J. H & Keaffaber J. J. Ozonation products of bromine and chlorine in seawater aquaria. Water Research 1994; 28 (5) 1087-1094.

[12] Zakaria P., Bloomfield C., Shellie R. A, Haddad P. R & Dicinoski G. W. Determination of bromate in sea water using multi-dimensional matrix-elimination ion chromatography. Journal of Chromatography A 2011; 1218(50) 9080-9085.

[13] Ritar A. J, Smith G. G & Thomas C. W. Ozonation of seawater improves the survival of larval southern rock lobster, Jasus edwardsii, in culture from egg to juvenile, Aquaculture 2006; 261(3) 1014-1025.

[14] Kumar S. D., Maiti B. & Mathur P. K. Determination of iodate and sulfate in iodized common salt by ion chromatography with conductivity detection. Talanta 2001; 53(4) 701-705.

[15] Rebary B., Paul P. & Ghosh P. K. Determination of iodide and iodate in edible salt by ion chromatography with integrated amperometric detection. Food Chemistry 2010; 123(2) 529-534.

[16] Richens D. A., Simpson D., Peterson S., Mcginn A. & Lamb J. D. Journal of Chromatography A 2003; 1016 (2) 155-164.

[17] Lamb J. D., Simpson D., Jensen B. D., Gardner J. S. & Peterson Q. P. Determination of perchlorate in drinking water by ion chromatography using macrocycle-based concentration and separation methods. Journal of Chromatography A 2006;1118(1) 100-105.

[18] Book of SEMI Standards (1995). Process Chemicals Volume, Semiconductor Equipment and Materials International, Mountain View, CA, 1995, 202.

[19] Report, T. R. PWR Secondary Water Chemistry Guidelines, Electric Power Research Institute, Palo Alto, CA, May (1993).

[20] Weiss J. Hand Book of Ion Chromatography, VCH, Weinheim 1995 (1) 360-367.

[21] Kaiser E., Riviello J., Rey M. Statler J. & Heberling S. Determination of trace level ions by high-volume direct-injection ion chromatography, Journal of Chromatography A (1996).

[22] Blazewicz A., Dolliver W., Sivsammye S., Deol A. & Randhawa, R. Orlic-zSzczesna G., Błazewicz R., Determination of cadmium, cobalt, copper, iron, manganese, and zinc in thyroid glands of patients with diagnosed nodular goitre using ion chromatography, Journal of Chromatography B 2010 (878) 34–38.

[23] Gautier C., Bourgeois M., Isnard H., Nonell A., Stadelmann G. & Goutelard F. Development of cadmium/silver/palladium separation by ion chromatography with quadrupole inductively coupled plasma mass spectrometry detection for off-line cadmium isotopic measurements, Journal of Chromatography A 2011; 1218 (31)5241-5247.

[24] Zhang Z., Xu Y., Wang C., Chen K., Tong H. & Liu S. Direct determination of hydrogen cyanide in cigarette mainstream smoke by ion chromatography with pulsed amperometric detection. Journal of Chromatography A 2011; 1218(7) 1016-1019.

[25] Zhao F., Wang Z., Wang H. & Ding M. Determination of hippuric acid in human urine by ion chromatography with conductivity detection. Journal of Chromatography B (2011).

[26] Waidyanatha S., Rothman N., Li G., Smith M. T., Yin S. & Rappaport S. M. Rapid determination of six urinary benzene metabolites in occupationally exposed and unexposed subjects. Analytical Biochemistry 2004; 327(2) 184-199.

[27] Erupe M. E., Liberman-martin A., Silva P. J., Malloy Q. G. J., Yonis N., Cocker D. R. & Purvis-Roberts K. L. Determination of methylamines and trimethylamine-N- oxide in particulate matter by non-suppressed ion chromatography. Journal of Chromatography A; 2010; 1217(13) 2070-2073.

[28] Moldoveanu S. C. & Kiser M. Gas chromatography/mass spectrometry versus liquid chromatography/fluorescence detection in the analysis of phenols in mainstream cigarette smoke Journal of Chromatography A 2007; 1141(1) 90-97.

[29] Jonsson G., Stokke T. U., Cavcic A., Jgensen, K. B. & Beyer J. Characterization of alkylphenol metabolites in fish bile by enzymatic treatment and HPLC-fluorescence analysis Chemoshpere 2008; 71(7) 1392-1400.

[30] Meyer J., Liesener A., Gotz S., Hayen H. & Karst U. Liquid Chromatography with On-Line Electrochemical Derivatization and Fluorescence Detection for the Determination of Phenols, Analalytical Chemistry. (2003).

[31] Wu S., Yang B., Xi L. & Zhu Y. Determination of phenols with ion chromatography-online electrochemical derivatization based on porous electrode-fluorescence detection, Journal of Chromatography A 2012; 1229, 288-292.

Chapter 2

Ion-Exchange Chromatography and Its Applications

Özlem Bahadir Acikara

Additional information is available at the end of the chapter

http://dx.doi.org/10.5772/55744

1. Introduction

Ion-exchange chromatography (IEC) is part of ion chromatography which is an important analytical technique for the separation and determination of ionic compounds, together with ion-partition/interaction and ion-exclusion chromatography [1]. Ion chromatography separation is based on ionic (or electrostatic) interactions between ionic and polar analytes, ions present in the eluent and ionic functional groups fixed to the chromatographic support. Two distinct mechanisms as follows; ion exchange due to competitive ionic binding (attraction) and ion exclusion due to repulsion between similarly charged analyte ions and the ions fixed on the chromatographic support, play a role in the separation in ion chromatography. Ion exchange has been the predominant form of ion chromatography to date [2]. This chromatography is one of the most important adsorption techniques used in the separation of peptides, proteins, nucleic acids and related biopolymers which are charged molecules in different molecular sizes and molecular nature [3-6]. The separation is based on the formation of ionic bonds between the charged groups of biomolecules and an ion-exchange gel/support carrying the opposite charge [7]. Biomolecules display different degrees of interaction with charged chromatography media due to their varying charge properties [8].

The earliest report of ion-exchange chromatography date back to 1850, Thompson studied the adsorption of ammonium ions to soils [9-11]. Spedding and Powell published a series of papers describing practical methods for preparative separation of the rare earths by displacement ion-exchange chromatography in 1947. Beginning in the 1950s, Kraus and Nelson reported numerous analytical methods which are used for metal ions based on separation of their chloride, fluoride, nitrate or sulfate complexes by anion chromatography [12]. In order to separate proteins an ion exchange chromatographic method was reported by Peterson and Sober in 1956. In modern form ion exchange chromatography was introduced by Small, Stevens and Bauman in 1975 [3]. Gjerde et al. published a method for anion chromatography in 1979 and this was followed by a similar method for cation chromatography in 1980 [12]. Ion-exchange chromatography has been used for many years to separate various ionic compounds; cations and anions and still continues to be used. The popularity of ion exchange chromatography has been increased in recent years because this technique allows analysis of wide range of molecules in pharmaceutical, biotechnology, environmental, agricultural and other industries [2].

1.1. Ion exchange mechanism

Ion-exchange chromatography which is designed specifically for the separation of differently charged or ionizable compounds comprises from mobile and stationary phases similar to other forms of column based liquid chromatography techniques [9-11]. Mobil phases consist an aqueous buffer system into which the mixture to be resolved. The stationary phase usually made from inert organic matrix chemically derivative with ionizable functional groups (fixed ions) which carry displaceable oppositely charged ion [11]. Ions which exist in a state of equilibrium between the mobile phase and stationary phases giving rise to two possible formats, anion and cation exchange are referred to as counter ion (Figure 1) [1,13]. Exchangeable matrix counter ions may include protons (H^+), hydroxide groups (OH^-), single charged mono atomic ions (Na^+, K^+, Cl^-), double charged mono atomic ions (Ca^{2+}, Mg^{2+}), and polyatomic inorganic ions (SO_4^{2-}, PO_4^{3-}) as well as organic bases (NR_2H^+) and acids (COO^-) [11]. Cations are separated on cation-exchange resin column and anions on an anion exchange resin column [10]. Separation based on the binding of analytes to positively or negatively charged groups which are fixed on a stationary phase and which are in equilibrium with free counter ions in the mobile phase according to differences in their net surface charge (Figure 1) [13-14].

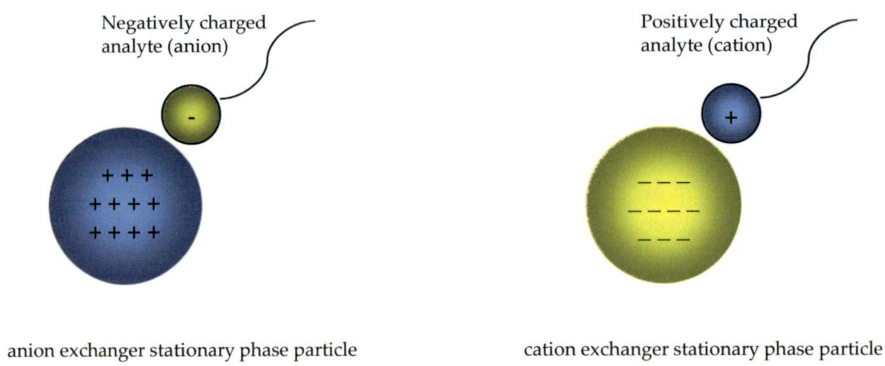

Figure 1. Types of ion exchangers

Ion exchange chromatography involves separation of ionic and polar analytes using chromatographic supports derivatized with ionic functional groups that have charges opposite that of the analyte ions. The analyte ions and similarly charged ions of the eluent compete to bind to the oppositely charged ionic functional group on the surface of the stationary phase. Assuming that the exchanging ions (analytes and ions in the mobile phase) are cations, the competition can be explained using the following equation;

$S\text{-}X^{-}C^{+} + M^{+} \leftrightarrow S\text{-}X^{-}M^{+} + C^{+}$

In this process the cation M^{+} of the eluent replaced with the analyte cation C^{+} bound to the anion X^{-} which is fixed on the surface of the chromatographic support (S).

In anion exchange chromatography, the exchanging ions are anions and the equation is represented as follow;

$S\text{-}X^{+}A^{-} + B^{-} \leftrightarrow S\text{-}X^{+}B^{-} + A^{-}$

The anion B^{-} of the eluent replaced with the analyte cation A^{-} bound to the positively charged ion X^{+} on the surface of the stationary phase. The adsorption of the analyte to the stationary phase and desorption by the eluent ions is repeated during their journey in the column, resulting in the separation due to ion-exchange [2].

Molecules vary considerably in their charge properties and will exhibit different degrees of interaction with charged chromatography support according to differences in their overall charge, charge density and surface charge distribution. Net surface charge of all molecules with ionizable groups is highly pH dependent [13]. Therefore pH of the mobile phase should be selected according to the net charge on a protein of interest within a mixture is opposite to that of matrix functional group, that it will displace the functional group counter ion and bind the matrix. On the other hand oppositely charged proteins will not be retained. Adsorbed protein analytes can be eluted by changing the mobile phase pH which effect the net charge of adsorbed protein, so its matrix binding capacity. Moreover increasing the concentration of a similarly charged species within the mobile phase can be resulted in elution of bound proteins. During ion exchange chromatography for example in anion exchange as illustrated in Figure 2, negatively charged protein analytes can be competitively displaced by the addition of negatively charged ions. The affinity of interaction between the salt ions and the functional groups will eventually exceed that the interaction exists between the protein charges and the functional groups, resulting in protein displacement and elution by increasing gradually the salt concentration in the mobile phase [11].

Complex mixtures of anions or cations can usually be separated and quantitative amounts of each ion measured in a relatively short time by ion exchange chromatography [10]. In classical ion-exchange chromatography separations have been performed in the open-column mode. Column which is loosely packed with stationary phase as small particles made of 1-2 cm diameter glass. The mobile phase or eluent contains the competing ion and is passed continuously into the column and percolates through it under gravity. Sample mixture is applied to the top of the column and allowed to pass into the bed of ion- exchange material. Eluent flow is then resumed and fractions of eluent are collected at regular intervals from the column outlet. Open column ion-exchange chromatography is very slow due to low eluent flow-rates. Increasing flow rate may result in deteriorated separation efficiency (Figure 3). In modern ion-exchange chromatography the usage of high efficiency ion exchange materials combined with flow-through detection have overcome of these challenges. Separations are performed on the column which is filled with ion-exchanger as particles in uniform size. The particles of ion-exchange material are generally very much smaller than those used for classical open column

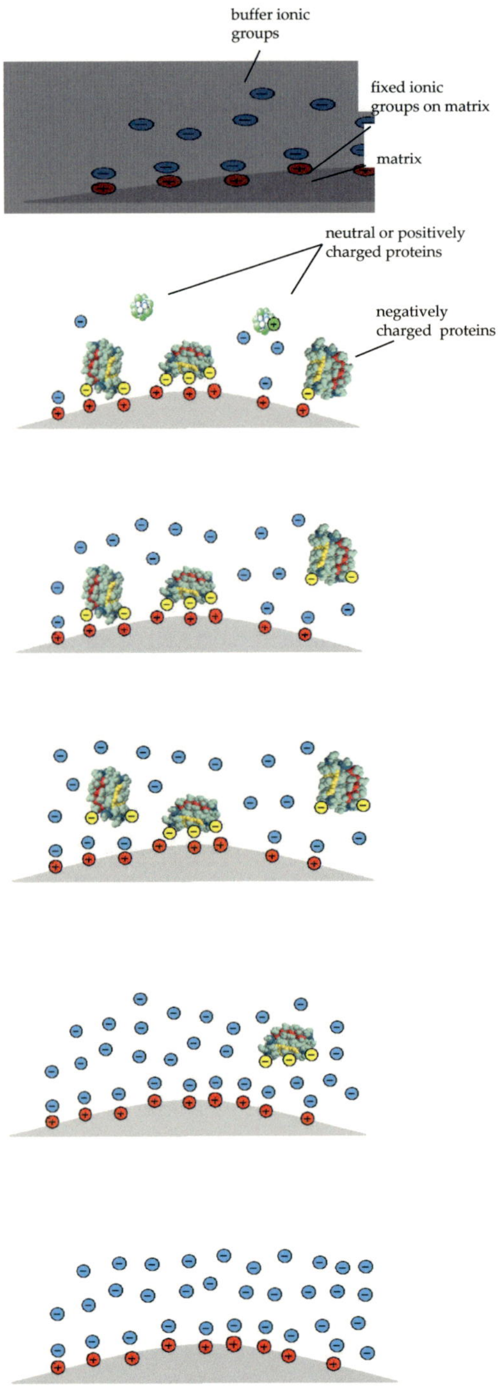

Equilibration
ion exchange medium equilibrated with starting buffer

Sample application
Oppositely charged proteins bind to ionic groups or to the ion exchange medium, becoming concentrated on the column. Uncharged proteins or those with the same charge as the ionic groups are eluted during or just after sample application

Elution
Increasing ionic strength (using a gradient) displaces bound proteins as ions in the buffer compete for binding sites

Elution
Further increases in ionic strength displace proteins that are more highly charged

Elution
Further increases in ionic strength displace proteins that are more highly charged

Washing
Final high ionic strength wash removes any ionically bound proteins before re-equilibration

Figure 2. Separation steps in anion exchange chromatography (GE Healthcare)

http://dx.doi.org/10.5772/55744

ion-exchange chromatography [1]. However ion-exchange resins used in modern chromatography have lower capacity than older resins [10]. The eluent must be pumped through the column due to the small particle size of stationary phase. The sample mixture is applied into eluent by the injection port. Finally the separated ions are detected with a flow-through detection instrument [1].

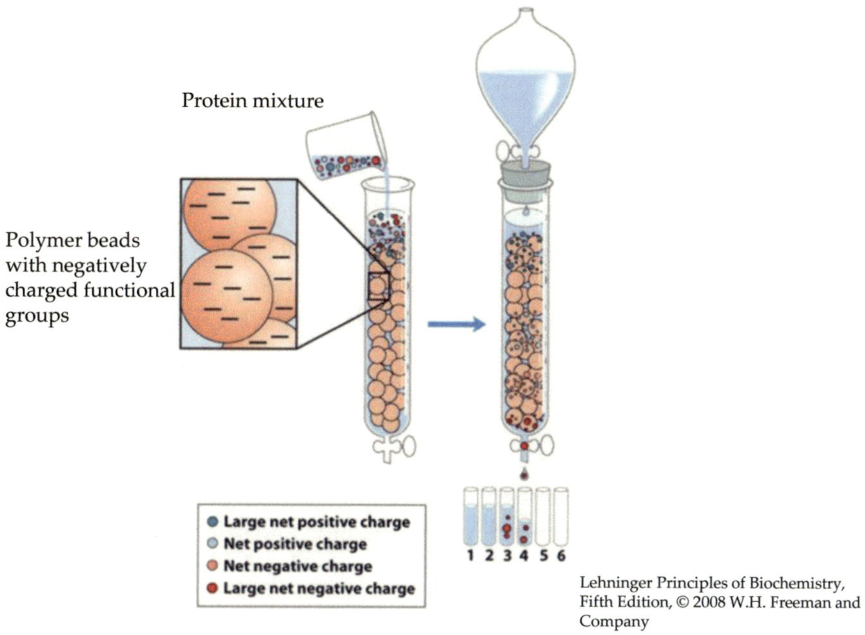

Figure 3. Ion exchange chromatography technique

This technique has been used for the analyses of anions and cations, including metal ions, mono- and oligosaccharides, alditols and other polyhydroxy compounds, aminoglycosides (antibiotics), amino acids and peptides, organic acids, amines, alcohols, phenols, thiols, nucleotides and nucleosides and other polar molecules. It has been successfully applied to the analysis of raw materials, bulk active ingredients, counter ions, impurities, and degradation products, excipients, diluents and at different stages of the production process as well as for the analysis of production equipment cleaning solutions, waste streams, container compatibility and other applications [2]. Wide applicability including high performance and high-throughput application formats, average cost, powerful resolving ability, large sample handling capacity and ease of scale-up as well as automation allow the ion exchange chromatography has become one of the most important and extensively used of all liquid chromatographic technique [11].

Although the extensive use of ion exchange chromatography the mechanism of the separation has not completely been elucidated. A considerable effort has been made to describe the ion exchange process theoretically [3,9]. One of the important disadvantages of this technique is that this method provides no direct information on events occurring at the surface of the stationary phase, because the ion-exchange equilibrium is always determined by the balance between the solute interaction and the eluent interaction with the active sites of resin [3]. Ion exchange is similar to sorption, since in both cases a solid takes up dissolved sample. The most important difference between them is in stoichiometric nature of ion exchange. Each ion removed from the solution is replaced by an equivalent amount of another ion of the same charge, while a solute is usually taken-up non-stoichiometrically without being replaced in sorption [15]. Stoichiometric displacement based on the mass action law and describes the retention of a solute ion as an exchange process with the counter ion bound to the surface [9]. According to this model, the retention of a protein under isocratic, linear conditions is related to counter ion concentration and can be represented by equilibrium as follow;

$$\log k = -(Z_p/Z_s) \log C_m + \log(\varphi Q)$$

k is the retention factor and C_m is the concentration of the counter ion in the mobile phase. Z_p/Z_s (= Z) is the ratio of the characteristic charge of the protein to the value of the counter ion and presents a statistical average of the electrostatic interactions of the protein with the stationary phase as it migrates through the column. The behavior of ion exchange chromatographic system can be explained by stoichiometric models. However, the mechanism of the ion exchange separation is more complex and stiochiometric consideration is inapplicable to long-range mechanisms, such as electrostatic interactions due to the distribution of ions in solution is also influenced by the electrostatic potential [3,6]. Other interactions between solute-solute, solute-solvent and solvent-solvent also contribute to retention and selectivity in ion exchange. For example ion-dipole and dispersion interaction, should be included as important mechanisms. Additionally entropic contribution originating from solvent, such as water, structures around ion exchange sites should also be regarded as important [3]. In addition to these the primary separation mechanism is the electrostatic interaction between ion-exchange sites and counter ions in ion exchange chromatography [6].

An important feature differentiating the ion exchange resins from other types of gels is the presence of functional groups. The groups are attached to the matrix. The ion exchange process between the ions in the solution takes place on these functional groups. The exchange of ions between the ion exchange resin and the solution is governed by two principles:

1. The process is reversible, only rare exceptions are known

2. The exchange reactions take place on the basis of equivalency in accordance with the principle of electro neutrality. The number of milimoles of an ion sorbed by an exchange should correspond to the number of milimoles of an equally charged ion that has been released from the ion exchange [16].

Equilibrium is established for each sample component between the eluent and stationary phases when a sample is introduced into the ion-exchange chromatography. The distribution of component (A) between the two phases is expressed by the distribution coefficient, "D_A".

$$D_A = [A]_r / [A]_m$$

The value of D_A is dependent on the size of the population of molecules of component A in the stationary and eluent phases [1]. As the equilibrium is dynamic, there is a continual, rapid interchange of molecules of component A between the two phases. The fraction of time, fm, that an average molecule of A spends in the mobile phase is given by:

f_m = Amount of A in the mobile phase / Total amount of A

$f_m = [A]_m V_m / [A]_m V_m + [A]_r w$

$= 1/ 1+ D_A (w / V_m)$

$k' = D_A (W / V_m)$

$f_m = 1 / 1 + k'$

w: Weight of the stationary phase

V_m: Volume of the mobile phase [1]

The mechanism of the anion and cation exchange are very similar. When analytes enter to the ion exchange column, firstly they bind to the oppositely charged ionic sites on the stationary phase through the Coulombic attraction [2]. In accordance with Coulomb's law, the interactions between ions in the solute and oppositely charged ligands on the matrix in ion-exchange chromatography are due to the electrostatic forces. Coulomb's law is given by the equation as follow;

$f = q_1 q_2 / \varepsilon r^2$

f: Interaction electrostatic force

$q_1 q_2$: The charge on ions

: Dielectric constant of the medium

r: The distance between charges.

If the charges on both ions are same (both are positive or negative) the force is repulsive, if they are different (one positive and the other negative) the force is attractive. When the ion charge of the species increase (Divalent ion should interact more strongly than a monovalent ion) and when the dielectric constant decrease (Two oppositely charged molecules increased more strongly in an organic solvent than in water), the interactions increase. On the other hand the distance between the charges increases the interactions decrease. Additionally, other interactions, especially, van der Waals forces participate to the Coulombic forces [2,17].

Ion exchange chromatography, which is also known as adsorption chromatography, is a useful and popular method due to its;

- high capacity,
- high resolving power,
- mild separation conditions,

- versatility and widespeared applicability,
- tendency to concentrate the sample
- relatively low cost [17].

General components of an ion-exchange chromatography are presented as below (Figure 4).

- A high pressure pump with pressure and flow indicator, to deliver the eluent
- An injector for introducing the sample into the eluent stream and onto the column
- A column, to separate the sample mixture into the individual components
- An oven, optional
- A detector, to measure the analyte peaks as eluent from the column
- A data system for collecting and organizing the chromatograms and data

In ion-exchange chromatography, adsorption and desorption processes are determined by the properties of the three interacting entities;

- The stationary phase,
- The constituents of the mobile phase
- The solute [18].

1.2. Stationary phase

Selection of a suitable ion-exchange matrix probably is the most important in ion exchange protocol and is based on various factors such as; ion exchanger charge/strength, linear flow rate/sample volume and sample properties [11]. In ion-exchange chromatography, numerous stationary phases are available from different manufacturers, which vary significantly in a number of chemical and physical properties [6,18]. Stationary phases comprised of two structural elements; the charged groups which are involved in the exchange process and the matrix on which the charged groups are fixed [18]. Ion exchangers are characterized both by the nature of the ionic species comprising the fixed ion and by the nature of the insoluble ion-exchange matrix itself [1].

Ion exchangers are called cation exchangers if they have negatively charged functional groups and possess exchangeable cations. Anion exchangers carry anions because of the positive charge of their fixed groups [15]. The charged groups determine the specifity and strength of protein binding by their polarity and density while the matrix determines the physical and chemical stability and the flow characteristics of the stationary phase and may be responsible for unspecific binding effects [18].

General structure (fibrous or beaded form), particle size and variation, pore structures and dimensions, surface chemistry (hydrophilic or hydrophobic), swelling characteristics of matrix are important factors which effect chromatographic resolution [11,18]. Porosity of ion exchange

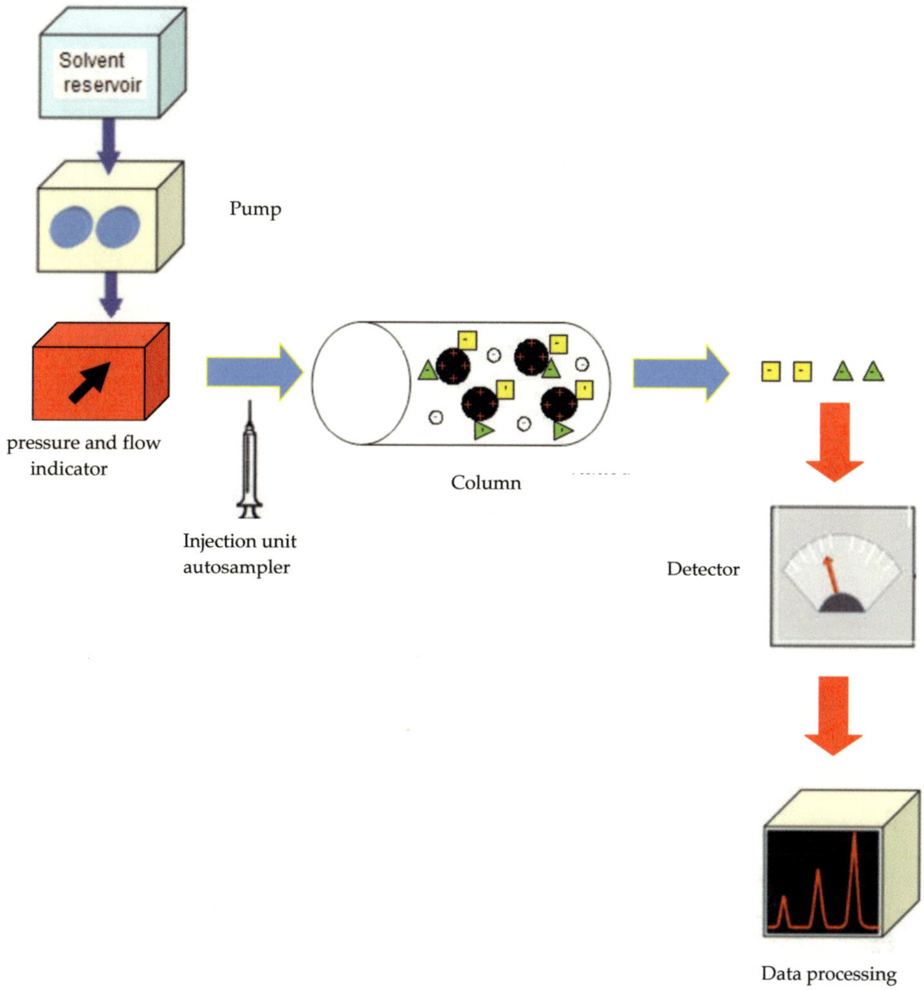

Figure 4. Ion-exchange Chromatography System

beads can be categorized as non-porous, microporous and macroporous. (Figure 5 and Figure 6) [14]. High porosity offers a large surface area covered by charged groups and so provides a high binding capacity [13]. However when compared with beaded matrix fibrous ion exchangers based on cellulose exhibit lower chromatographic resolution [14]. On the other hand high porosity is an advantage when separating large molecules [13] and prefractionation [14]. Non-porous matrices are preferable for high resolution separations when diffusion effects must be avoided [13]. Micropores increase the binding capacity but cause to a band broadening. Another disadvantage of microporous beads is that protein can bind to the surface of the beads near to the pores, so penetration of proteins into the pores can prevent or slow down. These problems are overcome by using macroporous particles with pore diameters of about 600-800 nm which are introduced recently. These kinds of particles behave differently compared to microporous materials with respect to microflow characteristics the new term perfusion chromatography has been created [14].

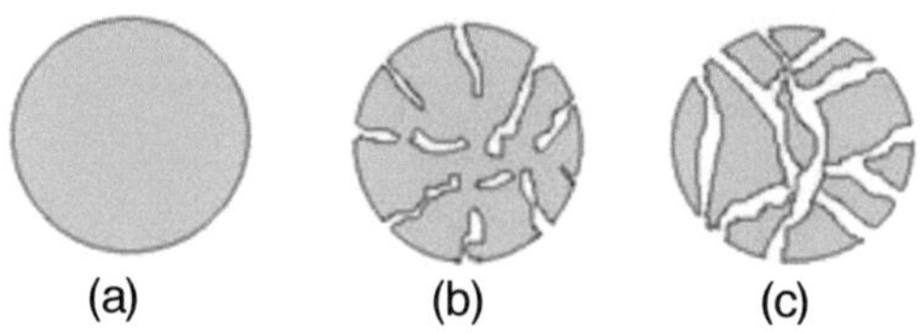

Figure 5. Schematic presentation of different matrix types (a) non-porous beads (b) microporous beads (c) macroporous beads

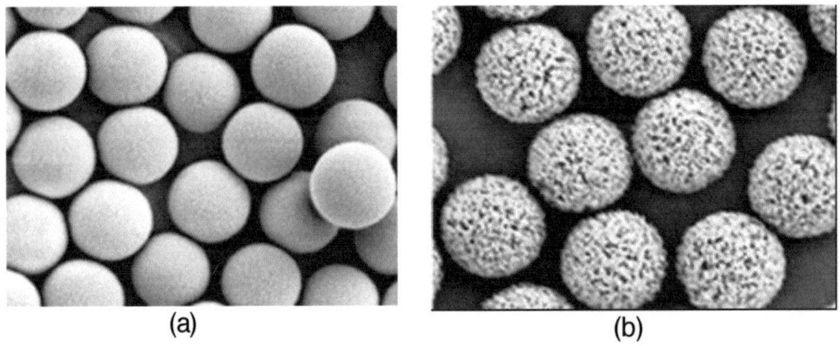

Figure 6. (a) Non-porous beads **(b)** Porous beads

Furthermore a new matrix type which has been recently introduced is based on a completely new principle and exhibits improved chromatographic features when compared with conventional ion exchangers. This matrix which is known as continuous bed does not consist of ion exchange beads or fibers. The matrix is synthesized in the column by polymerization and established from continuous porous support consisting of a nodule chains (Figure 7). The advantages of that matrix are mainly due to the more homogeneous mobile phase flow and short diffusion distances for the proteins. This is explained by the non-beaded form and the unique pore structure of the support [14].

Size, size distribution and porosity of the matrix particles are the main factors which affect the flow characteristics and chromatographic resolution. Small particles improved chromatographic resolution. Stationary phases with particle of uniform size are superior to heterogenous materials with respect to resolution and attainable flow rates. The pore size of ion exchange bead directly effect the binding capacity for a particular protein dependent on the molecular weight of the protein because it determines the access of proteins to the interior of the beads. Binding of large proteins can be restricted to the bead surface only so that the total binding capacity of the ion exchanger is not exploited Pore diameter of 30 nm is optimal for proteins up to a molecular weight of about 200.000 Da [14].

Figure 7. Continuous bed matrix

In order to minimize non-specific interactions with sample components inert matrix should be used. High physical stability provides that the volume of the packed medium remains constant despite extreme changes in salt concentration or pH for improving reproducibility and avoiding the need to repack columns. High physical stability and uniformity of particle size facilitate high flow rates, particularly during cleaning or re-equilibration steps, to improve throughput and productivity [13]. There are pH and pressure limits for each stationary phases. For example pH values higher than 8 should not used in silica based materials which are not coated with organic materials. Matrix stability also should be considered when the chemicals such as organic solvents or oxidizing agents should be required to use or when they are chosen for column cleaning [14].

Matrices which are obtained by polymerization of polystyrene with varying amounts of divinylbenzene are known as the original matrices for ion exchange chromatography. However these matrices have very hydrophobic surface and proteins are irreversibly damaged due to strong binding. Ion exchangers which are based on cellulose with hydrophilic back-bones are more suitable matrices for protein separations. Other ion exchange matrices with hydrophilic properties are based on agarose or dextran [14].

Several matrix types and their important properties can be listed as follow;

Matrix materials;

- Cellulose; Hydrophilic surface, enhanced stability by cross-linking, inexpensive

- Dextran; Considerable swelling as a function of ionic milieu, improved materials by cross-linking)
- Agarose; Swelling is almost independent of ionic strength and pH, high binding capacity obtained by production of highly porous particles
- Polyacrylamide; Swelling behavior similar to dextran
- Acrylate-copolymer; High pH stability
- Polystyrene-divinilybenzene; Hydrophobic surface, low binding capacity for proteins
- Coated polystyrene-divinilybenzene; Hydrophilic surface
- Silica; Unstable at pH > 8, rigid particles
- Coated Silica; Hydrophilic surface [14]

In addition to electrostatic interactions between stationary phase and proteins, some further mechanisms such as hydrophobic interactions, hydrogen bonding may contribute to protein binding. Hydrophobic interactions especially occur with synthetic resin ion exchangers such as which are produced by copolymerization of styrene and divinylbenzene. These materials are not usually used for separation of proteins. However new ion exchange materials that consist of styrene-divinylbenzene copolymer beads coated with hydrophilic ion exchanger film were introduced. According to the retention behavior of some proteins, it is considered that coating of the beads so efficient that unspecific binding due to hydrophobic interactions cannot be observed. Silica particles have also been coated with hydrophilic matrix. Acrylic acid polymers are also used for the protein separation in ion exchange chromatography. These polymers are especially suitable for purification of basic proteins [14].

The functional groups substituted onto a chromatographic matrix determine the charge of an ion exchange medium; positively-charged anion exchanger or a negatively-charged cation exchanger [13]. Both exchangers can be further classified as strong and weak type as shown in Table 1. The terms weak and strong are not related to the binding strength of a protein to the ion exchanger but describe the degree of its ionization as a function of pH [14]. Strong ion exchangers are completely ionized over a wide pH range, while weak ion exchangers are only partially ionized a narrow pH range [1,11]. Therefore with strong ion exchangers proteins can adsorb to several exchanger sites. For this reason strong ion exchangers are generally used for initial development and optimization of purification protocols. On the other hand weak ion exchangers are more flexible in terms of selectivity and are a more general option for the separation of proteins that retain their functionality over the pH 6-9 range as well as for unstable proteins that may require mild elution conditions [11]. Alkylated amino groups for anion exchangers and carboxy, sulfo as well as phosphato groups for cation exchangers are the most common functional groups used on ion exchange chromatography supports [14]. Sulfonic acid exchangers are known as strong acid type cation exchangers. Quaternary amine functional groups are the strong base exchangers whereas less substituted amines known as weak base exchangers [1]. Number and kind of the substituents are determined the basicity of amino-groups. Immobilized tertiary and quaternary amines proved to be useful for ion exchange chromatography. Immobilized diethylaminoethyl and carboxymethyl groups are the most widely used ion exchangers [11].

The ion exchange capacity of an ion-exchanger is determined by the number of functional groups per unit weight of the resin [13]. The total ionic capacity is the number of charged functional groups per ml medium, a fixed parameter of each medium and can be given as mval/ml for small ions. Density and accessibility of these charged groups both on the surface and within the pores define the total binding capacity. Ionic medium and the presence of other proteins if a particular protein is considered also affect the binding capacity. However, under defined conditions, the amount of the certain protein which is bound to ion exchanger is more suitable parameter for determining and comparing the capacity of ion exchange chromatography. Albumin for anion exchangers and hemoglobin for cation exchangers is usually used for this purpose. Determination of the binding capacity before the experiment is generally recommended because the capacity for a particular protein depends on its size and also on the sample composition. The binding capacity of a column can be increased for proteins which are retained on the column at high salt concentrations. The salt concentration is adjusted to a suitable concentration in which the protein of interest tightly bound to the ion exchanger while others which have lower affinity pass through the column without occupying binding sites [14].

Exchange Type	Ion exchange group	Buffer counter ions	pH range	Commercial samples
Strong cation	Sulfonic acid (SP)	Na^+, H^+, Li^+	4-13	Capto®S
				SP Sepharose®
				SP Sephadex®
				TSKgel SP_5PW
Weak cation	Carboxylic acid	Na^+, H^+, Li^+	6-10	CM Cellulose
				CM Sepharose®
				CM Sephadex®
				CM Sepharose® CL6B
				TSKgel CM-5PW
Strong anion	Quaternary amine (Q)	Cl^-, $HCOO_3^-$, CH_3COO^-, SO_4^{2-}	2-12	Q Sepharose®
				Capto®Q
				Dowex®1X2
				Amberlite® / Amberjet®
				QAE Sephadex®
Weak anion	Primary amine Secondary amine Tertiary amine (DEAE)	Cl^-, $HCOO_3^-$, CH_3COO^-, SO_4^{2-}	2-9	DEAE-Sepharose®
				Capto® DEAE
				DEAE Cellulose

Table 1. Weak and Strong type anion and cation exchangers

1.3. Mobile phase (Eluent)

In ion exchange chromatography generally eluents which consist of an aqueous solution of a suitable salt or mixtures of salts with a small percentage of an organic solvent are used in which most of the ionic compounds are dissolved better than in others in. Therefore the application of various samples is much easier [1,3]. Sodium chloride is probably the most widely used and mild eluent for protein separation due to has no important effect on protein structure. However NaCl is not always the best eluent for protein separation. Retention times, peak widths of eluted protein, so chromatographic resolution are affected by the nature of anions and cations used. These effects can be observed more clearly with anion exchangers as compared to cation exchangers [14]. The salt mixture can itself be a buffer or a separate buffer can be added to the eluent if required. The competing ion which has the function of eluting sample components through the column within reasonable time is the essential component of eluting sample. Nature and concentration of the competing ions and pH of the eluent are the most important properties affecting the elution characteristics of solute ions [1].

The eluent pH has considerable effects on the functional group which exist on the ion exchange matrix and also on the forms of both eluent and solute ions. The selectivity coefficient existing between the competing ion and a particular solute ion will determine the degree of that which competing ion can displace the solute ion from the stationary phase. As different competing ions will have different selectivity coefficients, it follows that the nature of competing ion will be an important factor in determining whether solute ions will be eluted readily. The concentration of competing ion exerts a significant effect by influencing the position of the equilibrium point for ion-exchange equilibrium. The higher concentration of the competing ion in the eluent is more effectively displace solute ions from the stationary phase, therefore solute is eluted more rapidly from the column. Additionally elution of the solute is influenced by the eluent flow-rate and the temperature. Faster flow rates cause to lower elution volumes because the solute ions have less opportunity to interact with the fixed ions. Temperature has relatively less impact, which can be change according to ion exchange material type. Enhancement of the temperature increases the rate of diffusion within the ion-exchange matrix, generally leading to increased interaction with the fixed ions and therefore larger elution volumes. At higher temperatures chromatographic efficiency is usually improved [1].

Eluent degassing is important due to trap in the check valve causing the prime loose of pump. Loss of prime results in erratic eluent flow or no flow at all. Sometimes only one pump head will lose its prime and the pressure will fluctuate in rhythm with the pump stroke. Another reason for removing dissolved air from the eluent is because air can get result in changes in the effective concentration of the eluent. Carbon dioxide from air dissolved in water forms of carbonic acid. Carbonic acid can change the effective concentration of a basic eluent including solutions of sodium hydroxide, bicarbonate and carbonate. Usually degassed water is used to prepare eluents and efforts should be made

http://dx.doi.org/10.5772/55744

to keep exposure of eluent to air to a minimum after preparation. Modern inline degassers are becoming quite popular [10].

For separation the eluent is pumped through the system until equilibrium is reached, as evidenced by a stable baseline. The time required for equilibrium may vary from a couple of minutes to an hour or longer, depending on the type of resin and eluent used [10]. Before the sample injection to the column should be equilibrated with eluent to cover all the exchange sites on the stationary phase with the same counter ion. When the column is equilibrated with a solution of competing ion, counter ions associated with the fixed ions being completely replaced with competing ions. In this condition the competing ions become the new counter ions at the ion exchange sites and the column is in the form of that particular ion [1].

Isocratic elution or gradient elution can be applied for elution procedure. A single buffer is used throughout the entire separation in isocratic elution. Sample components are loosely adsorbed to the column matrix. As each protein will have different distribution coefficient separation will achieved by its relative speeds of migration over the column. Therefore in order to obtain optimum resolution of sample components, a small sample volume and long exchanger column are necessary. This technique is time consuming and the desired protein invariably elutes in a large volume. However no gradient-forming apparatus is required and the column regeneration is needless. Alteration in the eluent composition is needed to achieve desorption of desired protein completely. To promote desired protein desorption continuous or stepwise variations in the ionic strength and/or pH of the eluent are provided with gradient elution. Continuous gradients generally give better resolution than stepwise gradients [11].

Additives which are protective agents found in the mobile phase are generally used for maintain structure and function of the proteins to be purified. This is achieved by stimulating an adequate microenvironment protection against oxidation or against enzymatic attacks [14]. Any additives used in ion exchange chromatography, should be checked for their charge properties at the working pH in order to avoid undesired effects due to adsorption and desorption processes during chromatography [13-14]. It is recommended to include in the elution buffer those additives in a suitable concentration which have been used for stabilization and solubilization of the sample. Otherwise precipitation may occur on the column during elution [14]. For example; zwitterionic additives such as betaine can prevent precipitation and can be used at high concentrations without interfering with the gradient elution. Detergents are generally useful for solubilization of proteins with low aqueous solubility. Anionic, cationic, zwitterionic and non-ionic (neutral) detergents can be used during ion exchange chromatography. Guanidine hydrochloride or urea, known as denaturing agents can be used for initial solubilization of a sample and during separation. However, they should use if there is a requirement. Guanidine is a charged molecule and therefore can participate to the ion exchange process in the same way as NaCl during separation process [13].

Commonly used eluent additives which have been successfully used in ion exchange chromatography can be given as follow;

- EDTA; Ethylenediamine tetraacetic acid
- Polyols; Glycerol, glucose, and saccharose
- Detergents;
- Urea and guanidinium chloride
- Lipids
- Organic solvents
- Zwitterions
- Sulfhydryl reagents
- Ligands
- Protease inhibitors [14]

1.4. Buffer

In ion exchange chromatography, pH value is an important parameter for separation and can be controlled and adjusted carefully by means of buffer substances [18]. In order to prevent variation in matrix and protein net charge, maintenance of a constant mobile phase pH during separation is essential to avoid pH changing which can occur when both protein and exchanger ions are released into the mobile phase [11]. By means of buffer substances pH value can be controlled and adjusted. Concentration of H^+ and the buffering component influence the protein binding to the stationary phase, chromatographic resolution and structural as well as functional integrity of the protein to be separated. Thus a suitable pH range, in which the stability of sample is guaranteed, has to be identified. Keeping of the sample function is related with the preservation of its three dimensional structure as well as with its biological activity [18]. A number of buffers are suitable for ion-exchange chromatography. A number of important factors influences the selection of mobile phase including buffer charge, buffer strength and buffer pH [11]. Properties of good buffers are high buffering capacity at the working pH, high solubility, high purity and low cost. The buffer salt should also provide a high buffering capacity without contributing much to the conductivity and should not interact with the ion exchanger functional groups as well as with media [11,17]. The buffering component should not interact with the ion exchanger because otherwise local pH shifts can occur during the exchange process which may interfere the elution. Interactions with stationary phase as well as with additives of the mobile phase and with subsequent procedures may be occur with buffer component and selected pH range. Precipitation of the mobile phase components can be observed for example when phosphate buffer and several di- and trivalent metal ions such as Mg^{+2} and Ca^{+2} are mixed or when anionic detergents (i.e. cholate) are used under acidic conditions or in the presence of multivalent metal ions. Precipitation of metal oxides and hydroxides can occur under alkaline conditions. Buffer components may also affect enzymatic assays used for screening and analysis of chromatography fractions [14]. The concentration of buffer salts usually ranges from 10 to 50 mM. Commonly used buffers are presented in Table 2 and Table 3 for cation and anion exchange chromatography [17].

Generally, applications of ion exchange chromatography are performed under slightly acidic or alkali conditions, pH range 6.0-8.5 but there are also more acidic and more alkali buffers. Additionally the buffering component should not act as an eluting ion by binding to the ion exchanger. Anionic buffer component such as phosphate or MOPS in cation exchange chromatography and cationic buffers such as ethanolamine, Tris and Tricine in anion exchange chromatography are recommended. Besides interactions of buffer component with stationary phase, there are also possible interactions with additives of the mobile phase. To achieve sufficient buffer capacity the pKa of the buffer component should be as close to the desired pH value as possible difference no more than ± 0.5 pH units. However there are examples of successful separations at which the buffering capacity is very low [17-18]. It has to be considered that the pKa is a temperature dependent value. Performing on ion exchange separation with the same elution buffer at room temperature or in the cold room can have a remarkable effect on the buffer capacity. For optimal binding of a sample ion to an ion-exchanger the ionic strength and thus also the buffer concentrations has to be low in sample and equilibration buffers [18].

Substance	pK_a	Working pH
Citric acid	3.1	2.6-3.6
Lactic acid	3.8	3.4-4.3
Acetic acid	4.74	4.3-5.2
2-(*N*-morpholino)ethanesulfonic acid	6.1	5.6-6.6
N-(2-acetamido)-2-iminodiacetic acid	6.6	6.1-7.1
3-(*N*-morpholino)propanesulfonic acid	7.2	6.7-7.7
Phosphate	7.2	6.8-7.6
N-(2-hydroxyethyl)piperazine-*N'*-(2-ethanesulfonic acid)	7.5	7.0-8.0
N,N-bis(2-hydroxyethyl)glycine	8.3	7.6-9.0

Table 2. Commonly used buffers for cation-exchange chromatography

Substance	pK_a	Working pH
N-Methyl-piperazine	4.75	4.25-5.25
Piperazine	5.68	5.2-6.2
Bis-Tris	6.5	6.0-7.0
Bis-Tris propane	6.8	6.3-7.3
Triethanolamine	7.8	7.25-8.25
Tris	8.1	7.6-8.6
N-Methyl-diethanolamine	8.5	8.0-9.0
Diethanolamine	8.9	8.4-9.4
Ethanolamine	9.5	9.0-10.0
1,3-Diaminopropane	10.5	10.0-11.0

Table 3. Commonly used buffers for anion-exchange chromatography

1.5. Detection

Conductivity detector is the most common and useful detector in ion exchange chromatography. However UV and other detectors can also be useful [10]. Conductivity detection gives excellent sensitivity when the conductance of the eluted solute ion is measured in an eluent of low background conductance. Therefore when conductivity detection is used dilute eluents should be preferred and in order for such eluents, to act as effective competing ions, the ion exchange capacity of the column should be low [1].

Although recorders and integrators are used in some older systems, generally in modern ion exchange chromatography results are stored in computer. Retention time and peak areas are the most useful information. Retention times are used to confirm the identity of the unknown peak by comparison with a standard. In order to calculate analyte concentration peak areas are compared with the standards which is in known concentration [10].

Direct detection of anions is possible, providing a detector is available that responds to some property of the sample ions. For example anions that absorb in the UV spectral region can be detected spectrophotometrically. In this case, an eluent anion is selected that does not absorb UV. The eluent used in anion chromatography contains an eluent anion, E^-. Anions with little or no absorbance in the UV spectral region can be detected spectrophotometrically by choosing a strongly absorbing eluent anion. An anion with benzene ring would be suitable [10]. Usually Na^+ or H^+ will be the cation associated with E^-. The eluent anion must be compatible with the detection method used. For conductivity the detection E should have either a significantly lower conductivity than the sample ions or be capable of being converted to a non-ionic form by a chemical suppression system. When a spectrophotometric detection is employed, E will often be chosen for its ability to absorb strongly in the UV or visible spectral region. The concentration of E^- in the eluent will depend on the properties of the ion exchanger used and on the types of anions to be separated [10].

2. Ion exchange chromatograpy applications

Ion exchange chromatography can be applied for the separation and purification of many charged or ionizable molecules such as proteins, peptides, enzymes, nucleotides, DNA, antibiotics, vitamins and etc. from natural sources or synthetic origin. Examples in which ion exchange chromatography was used as a liquid chromatograpic technique for separation or purification of bioactive molecules from natural sources can be given as below.

Sample 1:	
Source:	*Nigella sativa* Linn.
Extraction procedure:	Water extract of *N. sativa* was prepared, dried and powdered. Powder was dissolved in phosphate buffer saline (pH 6.4) and centrifuged at 10.000 rpm for 30 min at 4 ºC. The supernatant was collected as the soluble extract by removing the oily layer and unsoluble pellet. Protein concentration of the soluble extract was determined with Bradford method. Then proteins dialyzed against 0.05 M phosphate buffer (pH 6.4) using 3500 MW cut off dialyzing bags and centrifuged.
Stationary Phase:	XK50/30 column (5 x 15 cm) of DEAE sephadex A50.
Eluent:	0.05 M phosphate buffer (pH 6.4) containing 0.01 M NaCl. Fractions of each were collected with an increasing concentration of NaCl
Detection:	UV detector at 280 nm
Analyte(s):	Number of protein bands ranging from 94-10 kDa molecular mass [19].
Sample 2:	
Source:	*Olea europea* L.
Extraction procedure:	Extract was prepared from the leaves and roots of two years old olive plants with water at room temperature. Internal standard as D-3-*O*-methylglucopyranose (MeGlu) was used and added in appropriate volume. Extraction was accomplished by shaking for 15 min and finally the suspension was centrifuged at 3000 rpm for 10 min. Before the injection the aqueous phase was filtered and passed on a cartridge OnGuard A (Dionex) to remove anion contaminants.
Stationary Phase:	Two anion exchange columns Dionex CarboPac PA1 plus a guard column and CarboPac MA1 column with a guard column were used for separation procedure (High Performance Anion Exchange Chromatography).
Eluent:	Eluent was comprised 12 mM NaOH with 1 mM barium acetate. Flow rate was 1 mL/min.
Detection:	Pulsed amperometric detection
Analyte(s):	*myo*-inositol, galactinol, mannitol, galactose, glucose, fructose, sucrose, raffinose and stachyose [20].
Sample 3:	
Source:	Soybean
Extraction procedure:	Soybeans were defatted with petroleum ether for 30 min and centrifuged repeating the procedure twice. Then proteins were extracted with 0.03 M Tris-HCL buffer containing 0.01 M 2-mercaptoethanol (pH 8) for 1 hour following by centrifugation (16.250 x g for 20 min at 20 ºC). The supernatant was adjusted to pH 6.4 with 2 M HCl and centrifuged (16.250 x g for 20 min at 2-5 ºC). The precipitate was dissolved in Tris-HCl buffer and the process was repeated in order to obtain purified precipitated fraction containing the 11S globulin. The supernatant obtained after the first

precipitation of the 11S fraction was adjusted to pH 4.8 with 2M HCl and centrifuged (16.250 x g for 20 min at 2-5 °C). The supernatant was stored at low temperature and the precipitate was dissolved in Tris-HCl buffer (pH 8). The process was repeated to obtain a purified precipitated fraction containing the 7S globulin.

Stationary Phase: Anion exchange perfusion column POROS HQ/10 packed with cross-linked polystyrene-divinylbenzene beads.

Eluent: The starting point for the separation of soybean proteins by HPIEC was the use of a binary gradient where mobile phase was a buffer solution at a certain pH (always pHs higher than the isoelectric pH of soybean proteins, p*I* = 4.8–6.4) and mobile phase B was the same buffer solution containing as well Msodium chloride. The buffer solutions tried were: phosphate ffer (pH 7 and 12), Tris–HCl buffer (pH 8), borate buffer (pH 9), and carbonate buffer (pH 10). In all cases, the buffer concentration was 20 mM. For every buffer, different gradients were tried. The best separation for ybean proteins was obtained with the borate buffer (pH 9) and gradient starting with an isocratic step at 0% B for 2.5 min and from 0 to 70% B in 14 min (gradient slope, 5%B/min). A fine optimization of the selected gradient enabled a reduction of the analysis time keeping the separation. The final gradient was: 0% for 2 min and from 0 to 50% B in 10 min.

Detection: UV detector at 254 nm

Analyte(s): 11S globulin or glycinin and 7S globulin [21].

Sample 4:

Source: *Cochlospermum tinctorium* A. Rich.

Extraction procedure: The powdered roots of *C. tinctorium* were extracted with ethanol (% 96, v/v) using a soxhlet apparatus to remove low molecular weight compounds. Extraction procedures continue until no color could be observed in the ethanol. The residue was extracted with water at 50 °C, 2 hour for two times. Obtained extract was filtered through gauze and Whatman GF/A glass fiber filter and then concentrated at 40 °C in vacuum and dialysed at cut-off 3500 Da to give a 50 °C crude extract. The extracts was kept at -18 °C or lyophilized.

Stationary Phase: Anion exchange-DEAE-Sepharose column

Eluent: For obtaining neutral fraction the column was eluted with water firstly. The acidic fractions were obtained by elution of linear NaCl gradient (0-1.4 M) in water. The carbohydrate elution profile was determined using the phenol-sulphiric acid method. Finally two column volumes of a 2 M sodium chloride solution in water were eluted to obtain the most acidic polysaccharide fraction. The relevant fractions based on the carbohydrate profile were collected, dialysed and lyophilized.

Detection: UV detector, 490 nm

Analyte(s): Glucose, galactose, arabinose (in neutral fraction) Uronic acids (Both galacturonic and glucuronic acid), rhamnose, galactose, arabinose and glucose (in acidic fraction) [22].

Sample 5:

Source: Hen egg

Extraction procedure: Fresh eggs were collected and the same day extract was obtained. Ovomucin was obtained using isoelectric precipitation of egg white in the presence of 100 mM NaCl solution. The dispersion was kept overnight at 4 °C and separated by centrifugation at 15.300 x g for 10 min at 4 °C. The precipitate was further suspended in 500 mM NaCl solution while stirring for 4 h followed by overnight settling at 4 °C. After centrifugation at 15.300 x g for 10 min at 4 °C, the precipitate was freeze dried and stored at -20 °C. The supernatants obtained during the first step (with 100 mM NaCl solution) and the second step (with 500 mM NaCl solution) was further used for ion exchange chromatography to separate other egg white proteins. Separation proteins from 100 mM supernatant were allowed to pass through an anion exchange chromatographic column to separate different fractions. The unbound fractions were then passed through a cation exchange chromatographic column to separate further.

Stationary Phase: High-Prep 16/10 column (Q Sepharose FF)-Anion Exchange Chromatography High-Prep 16/10 column (SP Sepharose FF)-Cation Exchange Chromatography

Eluent: The column was equilibrated with water and the pH was adjusted to 8.0 before injectionAfter sample injection flow-through fraction was collected using water as the eluent, followed by isocratic elution of the sample using 0.14 M NaCl. Finally the bound fraction was eluted using gradient elution (0.14-0.5 M) of NaCl- Anion Exchange Chromatography.

The unbound fraction was collected and used as starting material for cation exchange chromatography. The column was equilibrated with 10 mM citrate buffer, which was used as the starting buffer. After sample injection the column was eluted by isocratic elution using 0.14 M NaCl solution followed by gradient elution from 0.14 M to 0.50 M NaCl solution. The fractions were collected and freeze dried-Cation Exchange Chromatography.

Detection: MS Detector

Analyte(s): Ovalbumin, ovotransferrin, lysozyme, ovomucin [23].

Sample 6:

Source : *Phaseolus vulgaris*

Extraction procedure: Seeds were grounded and soaked in 20 mM Tris-HCl buffer (pH 7.6) at 4 ºC for 24 h. The seeds were blended in a blender to extract the proteins followed by centrifugation (30,000g) at 4ºC. Then 450 g/l of ammonium sulphate were added to the supernatant to 70% saturation. The precipitate was removed by centrifugation and the supernatant was extensively dialysed against distiled water. The dialysed protein extract was freeze dried and used for chromatographic separation.

Stationary Phase: Q-Sepharose Column (3 cm x 7 cm), anion-exchange

Eluent: The column was equilibrated and initially eluted with 20 mM Tris–HCl (pH 7.6). Elution of the bound fraction was carried out by using 1 M NaCl in the equilibration buffer. All chromatographic steps were performed at the flow rate of 100 ml/h. Further separation selected fraction Q1, which was lyophilised and dissolved in 100 mM Tris–HCl (pH 7.6) buffer was performed onto a FPLC Superdex 75 column at a flow rate 0.5 ml min^{-1}.

Detection: UV detector, 280 nm

Analyte(s): A 5447 Da antifungal peptide [24].

Sample 7:

Source: Sweet dairy whey

Extraction procedure: After the cheese making process the sweet whey is produced, it is further processed by reverse osmosis to increase the solids content from approximately 5.5% (w/w) to 14.6% (w/w).

Stationary phase: Pharmacia's Q- and S-Sepharose anion- and cation-exchange resins

Eluent 1: For the anion-exchange process; it was found that two step changes, simultaneous in pH and salt concentration were necessary to carry out the anion-exchange separation. A 0.01 *M* sodium acetate buffer, pH 5.8, was used for the starting state or feed loading buffer. After the whey feed was loaded onto the column, one column volume of this buffer was passed through to wash out any material that did not bind to the resin, including the IgG. Next, two column volumes of 0.05 *M* sodium acetate, pH 5.0, were passed through the column to desorb those proteins whose p*I* values were above 5.0. This includes the β-lactoglobulin and bovine serum albumin. This was then followed by two column volumes of 0.1 *M* sodium acetate, pH 4.0, to finally desorb the α-lactalbumin whose p*I* range is 4.2–4.5, and thus above that of the passing pH wave of 4.0. After this second step change, the cleaning cycle was then implemented to prepare the column for the next run.

Eluent 2: For the cation-exchange process, it was found that one step change in pH was appropriate to carry out the cation-exchange separation. The buffer used was 0.05 *M* sodium acetate, pH 5.5, as the starting state or feed loading buffer. One column volume loading of the anion-exchange breakthrough curve fraction was optimum for loading onto the cation-exchange column. After the anion-exchange breakthrough curve fraction was loaded onto the column, one column volume of the initial buffer was passed through to wash out any material that did not bind to the resin. Next a step change in pH was implemented to elute the bound IgG. This was accomplished by passing two column volumes of the buffer, 0.05 *M* sodium acetate, pH 8.5. As the pH wave of this buffer passed through the cation bed it initiated the elution of the IgG because the upper value of its p*I* range is 8.3. After this pH step change the cleaning cycle was then implemented.

Detection: UV Detector

Analyte(s): α-lactalbumin, β-lactoglobulin, bovine serum albumin, immunoglobulin G and lactose [25].

Sample 8:

Source: *Morus alba* (mulberry) leaves

Extraction procedure: Fresh leaves were homogenized in ice-cold 50 mM Tris–HCl, pH 7.5, containing 0.3 M NaCl, 20 mM diethyldithiocarbamic acid, 5% glycerol, and 2% polyvinylpyrrolidone. The buffer used was 3 ml g^{-1} of the fresh leaves. The homogenate was filtered through a layer of cheesecloth and stored at 20°C for 24 h. After thawing, it was centrifuged at 8000x*g*, for 40 min at 4°C. The supernatant was collected and ammonium sulfate was added to 70% saturation. The resulting precipitate was recovered by centrifugation at 8000x*g* for 40 min, redissolved in tris-buffered saline, TBS (50 mM Tris–HCl, pH 7.5 containing 0.3 M NaCl) and dialysed against the buffer overnight at 4°C. The solution was then centrifuged at 13,000x*g* for 15 min and the supernatant was collected and stored at -20°C. An aliquot of the dialysed ammonium sulfate fraction containing protein was applied to the affinity chromatography on the *N*-acetylgalactosamine-agarose column. And then further separation was performed on Sephacryl S-200 column followed by anion exchange chromatography. Further purification was also performed by anion exchange and gel filtration chromatography

Stationary Phase: Anion-exchange chromatography, a DEAE-Sephacel column (2x9 cm)

Eluent: Equilibrated with 20 mM Tris–HCl, pH 7.5 at flow rate 20 ml min^{-1} and then eluted stepwise with the buffer containing NaCl.

Detection: UV Detector, 280 nm

Analyte(s): Lectins, MLL 1 and MLL 2 [26]

Sample 9:

Source: *Lycium ruthenicum* Murr.

Extraction: Fruits of the plant extracted with hot water yielded a crude polysaccharide sample, CLRP. The carbohydrate of CLRP was 66.2% and protein content was 7.3%. CLRP was a black Polysaccharide sample in which the pigment could not be removed by colum chromatography. To avoid the influence of pigment on the structure analysis, decoloration was performed with 30% H_2O. After decoloration, the carbohydrate content of decolored CLRP was 93.2% and protein content was 4.4%. Decolored CLRP was purified by anion exchange chromatography, yielding five polysaccharide subfractions LRP1, LRP2, LRP3, LRP4, and LRP5.

Stationary Phase: DEAE-cellulose column

Eluent: Distilled water, 0.05–0.50 mol/L $NaHCO_3^-$ solution

Detection: UV Detector, 280 nm

Analyte(s): Glycoconjugate polysaccharide (LRGP1) [27]

Sample 10:

Source: *Coprinus comatus*

Extraction: Stipe powder of *C. comatus* (100 g) was extracted three times with 1 L 95% ethanol under reflux for 2 h to remove lipid, and the residue was extracted three times with 2 L distilled water for 2 h at 80 °C with intermediate centrifugation (2000 × *g*, 15 min). After concentrating the collected aqueous supernatants to 400 mL (reduced pressure at 40 °C), a precipitation was performed with 3 volumes of 95% ethanol. The precipitate was washed with ethanol and acetone, and then dried at 40 °C, yielding crude polysaccharide material. Crude polysaccharide material was dissolved in 100 mL 0.2 M sodium phosphate buffer (pH 6.0), and after centrifugation the solution was applied to a DEAE-Sepharose CL-6B column.

Stationary Phase: DEAE-Sepharose CL-6B column (3.5 cm × 30 cm).

Eluent: 0.2 M sodium phosphate buffer (pH 6.0), and linear gradient of 0.3–1.5 M NaCl in 0.2 M sodium phosphate buffer (pH 6.0).

Detection: UV Detector, 490 nm (phenol–H_2SO_4) and 500 nm (Folin–phenol)

Analyte(s): Polysaccharides; disaccharide α,α-trehalose, α-D-glucan, β-D-glucan, α-L-fuco-α-D-galactan [28].

Sample 11:

Source: *Physalisalkekengi* var. *francheti*

Extraction: The dried and defatted fruit calyx extracted with different enzyme Neutral proteinase, Papain and alkaline protease, respectively, in their suitable pH and temperature and then each extract was centrifuged at 5000 rpm for 10 min. The supernatant was concentrated and then precipitated by the addition of ethanol in 1:4 (v/v) at room temperature. The precipitate was dissolved in distilled water and the solution was then washed with sevag reagent (isoamyl alcohol and chloroform in 1:4 ratio), which were centrifuged at 5000 rpm for 15 min and the protein was removed. The supernatant was dialyzed against deionized water for 24 h before concentration under vacuum evaporator at 55 °C. The mixture was precipitated by the addition of ethanol in 1:4 (v/v) at room temperature and the precipitate was freeze dried. Total sugars were determined by the phenol–sulfuric acid assay using glucose as standard.

Stationary Phase: DEAE anion-exchange column

Eluent: The column was eluted first with distilled water, and then with gradient solutions (0.1 M, 0.25 M, 0.5 M NaCl and 0.5 M NaOH), at a flow rate of 0.6 mL/min. The major polysaccharidefractions were collected with a fraction collector and concentrated using a rotary evaporator at 55 °C and residues were loaded onto a Sephadex G-200

	gel column (2.5 × 65 cm). The column was eluted with 0.1 M NaCl at a flow rate of 0.3 mL/min. The major fraction was collected and then freeze dried. All of these fractions were assayed for sugar content by the phenol–sulfuric acid method using glucose as standard
Detection:	UV Detector, 490 nm
Analyte(s):	Polysaccharides [29].
Sample 12:	
Source:	*Ornithogalum caudatum* Ait.
Extraction:	The whole dried plant was soaked with 95% ethanol to remove the pigments, defats and inactivates enzymes, and refluxed by hot distilled water for 4 h at 90 °C. The aqueous extract was concentrated to 30% of the original volume under reduced pressure in a rotary evaporator, and proteins were removed with Sevag method. The obtained solution was precipitated with 40% ethanol. The supernatant was added by ethanol up to 60%, and kept at 4 °C overnight. The polysaccharide pellets were obtained by centrifugation at 4000 rpm for 15 min, and completely dissolved in appropriate volume of distilled water followed by intensive dialysis for 2 days against distilled water (cut-off M_w 3500 Da). The retentate portion was then concentrated, and centrifuged to remove insoluble material. Finally the supernatant was lyophilized to give crude extract. The crude extract was dissolved in 0.2 mol/L tris (hydroxymethyl) aminomethane hydrochloride buffer solution, and filtered through a filter paper. The solution was passed through an anion-exchange chromatography column. After ion exchange chromatography other chromatographic methods was used for further separations.
Stationary Phase:	DEAE-Sepharose fast flow anion-exchange chromatography column (10 × 300 mm)
Eluent:	The polysaccharides were eluted with Tris–HCl buffer solution, followed with gradient elution of 0.1–0.8 mol/L NaCl at a flow rate of 0.8 ml/min.
Detection:	UV Detector, 486 nm (phenol–sulfuric acid method)
Analyte(s):	Water soluble polysaccharides [30].
Sample 13	
Source:	*Paecilomyces variotii*
Extraction:	After fermentation process, ammonium sulphate was added to the supernatant to give a final concentration of 80% saturation. The ammonium sulphate was added with constant stirring at 4ºC and the mixture stood overnight at 4ºC. The precipitated proteins were separated by centrifugation at 10000 rpm at 5ºC for 30 min. The separated proteins were then re-suspended in a minimum amount of distilled water and the solution dialyzed (using cellulose dialysis tubing) for 24 hrs against distilled water and concentrated by freeze-drying. The partially purified enzyme was dissolved in acetate buffer (20 mM - pH 6.0) and passed through a column
Stationary Phase:	Diethylaminoethyl (DEAE) Sepharose column (0.7 x 2.5 cm)
Eluent:	Acetate buffer (20 mM - pH 6.0)equilibrated with the same buffer. The solution was passed through the column at a flow rate of 1 $mL.mim^{-1}$ with acetate buffer (20 mM - pH 6.0), followed by a linear gradient from 0-1M NaCl in the acetate buffer. The eluted fractions were collected in an automated fraction collector (Pharmacia Biotech) and the absorbance of the fractions was measured at 280 nm. The major peak fractions were then assayed for tannase activity, and only the fractions possessing tannase activity were pooled.
Detection:	UV Detector, 280 nm

Analyte:	Tannase [31]
Sample 14	
Source:	*Castanospermum australe*
Extraction:	50% MeOH extract of seeds
Stationary Phase:	(1) Amberlite IR-120B (500 mL H^+ form), (2) Dowex 1-X2 column (3.8×90 cm, OH^- form), (3) Amberlite CG-50 column (3.8×90 cm, NH4 + form), (4) Dowex 1-X2 column (3.8×90 cm, OH^- form) (Repeated separation on different ion exchange columns).
Eluent:	0.5 M NH_4OH, H_2O
Detection:	UV Detection by HPTLC
Analyte(s):	Pyrolizidine alkaloids; fagomine; 6-epi-castanospermine; castanospermine; australine; 3-epi-fagomine; 2,3-diepi-australine; 2,3,7-triepi-australine; 3-epi-australine; 2R-hydroxymethyl-3S-hydroxypyrrolidine; castanospermine-8-O--D-glucopyranoside; 1-epi-australine-2-O--D-glucopyranoside and 1-epi-australine [32].

3. Conclusion

Since the isolation of pharmacologically active substances which are responsible for the activity became possible at the beginning of the 19th century drug discovery researches have increased dramatically [33]. Therefore within the last decade there has also been increasing interest in the liquid chromatographic processes because of the growing pharmaceutical industry and needs from the pharmaceutical and specialty chemical industries for highly specific and efficient separation methods. Several different types of liquid chromatography techniques are utilized for isolation of bioactive molecules from different sources [25]. Ion exchange chromatography is probably the most powerful and classic type of liquid chromatography. It is still widely used today for the analysis and separation of molecules which are differently charged or ionizable such as proteins, enzymes, peptides, amino acids, nucleic acids, carbohydrates, polysaccharides, lectins by itself or in combination with other chromatographic techniques [34]. Additionally ion exchange chromatography can be applied for separation and purification of organic molecules from natural sources which are protonated bases such as alkaloids, or deprotonated acids such as fatty acids or amino acid derivatives [35]. Ion exchange chromatography has many advantages. This method is widely applicable to the analysis of a large number of molecules with high capacity. The technique is easily transferred to the manufacturing scales with low cost. High levels of purification of the desired molecule can be achieved by ion exchange step. Follow-up of the nonsolvent extractable natural products can be realized by this technique [17,35]. Consequently ion exchange chromatography, which has been used in the separation of ionic molecules for more than half a century is still used as an useful and popular method for isolation of natural products in modern drug discovery and it continue to expand with development of new technologies.

Author details

Özlem Bahadir Acikara

Ankara University, Faculty of Pharmacy, Department of Pharmacognosy, Ankara, Turkey

References

[1] Haddad PR, Jackson PE. Journal of Chromatography Library-Volume 46 Ion Chromatography. Amsterdam: Elsevier Science; 1990.

[2] Bhattacharyya L, Rohrer JS. Applications of Ion Chromatography for Pharmaceutical and Biological Products. New Jersey: John Wiley & Sons; 2012.

[3] Okada T. Nonaqueous ion-exchange chromatography and electrophoresis Approaches to nonaqueous solution chemistry and design of novel separation. Journal of Chromatography A 1998; 804 17-28.

[4] Levison PR. Large-scale ion exchange column chromatography of proteins Comparison of different formats. Journal of Chromatography B 2003;790 17-33.

[5] Ishihara T, Kadoya T, Yamamoto S. Application of a chromatography model with linear gradient elution experimental data to the rapid sacle-up in ion-exchange process chromatography of proteins. Journal of Chromatography A 2007;1162 34-40.

[6] Bruch T, Graalfs H, Jacob L, Frech C. Influence of surface modification on protein retention in ion-exchange chromatography Evaluation using different retention models. Journal of Chromatography A 2009;1216 919-926.

[7] Stanton P. HPLC of Peptides and Proteins. Methods in Molecular Biology. New Jersey: Humana Press; 2004.

[8] Grodzki AC and Berenstein E. Antibody Purification:Ion-Exchange Chromatography. Immunocytochemical Methods and Protocols. Methods in Molecular Biology 2010;588 27-32.

[9] Wisel A, Schmidt-Traub H, Lenz J, Strube J. Modelling gradient elution of bioactive multicomponent systems in non-linear ion-excahnge chromatography. Journal of Chromatography A 2003;1006 101-120.

[10] Fritz JS, Gjerde DT. Ion Chromatography (Fourth Completely Revised and Enlarged Edition). Weinhein: Wiley-VCH Verlag GmbH & KGoA Weinhein; 2009.

[11] Cummins PM, Dowling O, O'Connor BF. Ion-Exchange Chromatography: Basic Principles and Application to the Partial Purification of Soluble Mammalian Prolyl Oligopeptides. In: Walls D, Loughran ST. (ed.) Protein Chromatography Methods and Protocols. New York: Springer; 2011. p215-228.

[12] Fritz JJ. Early milestones in the development of ion-exchange chromatography: a personal account. Journal of Chromatography A 2004;1039 3-12.

[13] GE Healthcare. Ion Exchange Chromatography & Chromatofocusing Principles and Methods Handbook. https://www.gelifesciences.com/gehcls_images/GELS/Related%20Content/Files/1314823637792/litdoc11000421AB_20110901010317.pdf (acessed 20 July 2012).

[14] Kastner M. Protein Liquid Chromatography. Amsterdam: Elsevier Science; 2000.

[15] Zagorodni AA. Ion Exchange Materials, Properties and Publications. Amsterdam: Elsevier Science; 2007.

[16] Korkisch J. Handbook of Ion Exchange Resins: Their Application to Inorganic Analytical Chemistry Volume V. Florida: CRC Press; 2000.

[17] Westerlund B. Ion-exchange Chromatography In Simpson RJ. Purifying Proteins for Proteomics. New York: Cold Spring Harbor Laboratory Press; 2004.

[18] Kastner M. Protein Liquid Chromatography. Amsterdam: Elsevier Science; 2005.

[19] Haq A, Lobo PI, Al-Tufail M, Rama NR, Al-Sedairy ST. Immunomodulatory effect of *Nigella sativa* proteins fractionated by ion exchange chromatography. International Journal of Immunopharmacology 1999;21 283-295.

[20] Cataldi TRI, Margiotta G, Iasi L, Di Chio B, Xiloyannis C, Bufo SA. Determination of Sugar Compounds in Olive Plant Exracts by Anion-Exchange Chromatography with Pulsed Amperometric Detection. Analytical Chemistry 2000;72 3902-3907.

[21] Heras JM, Marina ML, Garcia MC. Development of a perfusion ion exchange chromatography method for the separation of soybean proteins and its application to cultivar characterization. Journal of Chromatography A 2007;1153 97-103.

[22] Nergard CS, Diallo D, Inngjerdingen K, Michaelsen TE, Matsumoto T, Kiyohara H, Yamada H, Paulsen BS. Medicinal use of *Cochlospermum tinctorium* in Mali Anti-ulcer, radical scavenging and immunomodulating activities of polymers in the aqueous extract of the roots. Journal of Ethnopharmacology 2005; 69 255-269.

[23] Omana DA, Wang J, Wu J. Co-extraction of egg white proteins using ion-exchange chromatography from ovomucin-removed egg whites. Journal of Chromatography B 2010; 878 1771-1776.

[24] Wong JH, Ip DCW, NG TB, Chan YS, Fang F, Pan WL. A defensin-like peptide from Phaseolus vulgaris cv. 'King Pole Bean'. Food Chemistry 2012;135 408-414.

[25] Gerberding SJ, Byers CH. Prparative ion-exchange chromatography of proteins from dairy whey. Journal of Chromatography A 1998;808 141-151.

[26] Ratanapo S, Ngamjunyaporn W, Chulavatnatol M. Sialic acid binding lectins from leaf of mulberry (Morus alba). Plant Science 1998;139 141-148.

[27] Peng Q, Lv X, Xu Q, Li Y, Huang L, Du Y. Isolation and structural characterization of the polysaccharide LRGP1 from Lycium ruthenicum. Carbohydrate Polymers 2012;90 95-101.

[28] Li B, Dobruchowska JM, Gerwig GJ, Dijkhuizen L, Kamerling JP. Structural investigation of water-soluble polysaccharides extracted from the fruit bodies of Coprinus comatus. Carbohydrate Polymers 2013;91 314-321.

[29] Ge Y, Duan Y, Fang G, Zhang Y, Wang S. Polysaccharides from fruit calyx of Physalis alkekengi var. francheti Isolation, purification, structural features and antioxidant activities. Carbohydrate Polymers 2009;77 188-193.

[30] Chen R, meng F, Liu Z, Chen R, Zhang M. Antitumor activities of different fractions of polysaccharide purified from *Ornithogalum caudatum* Ait. Carbohydrate Polymers 2010;80 845-851.

[31] Battestin V, Macedo GA. Effects of temperature, pH and additives on the activity of tannase produced by Paecilomyces variottii. Electronic Journal of Biotechnology 2007;10 http://www.ejbiotechnology.cl/ content/vol10/issue2/full/9/index.html#6 (accesses 20.09.2012).

[32] Kato A, Kano E, Adachi I, Molyneux RJ, Watson AA, Nash RJ, Fleet GWJ, Wormald MR, Kizu H, Ikeda K, Asano N. Australine and related alkaloids: easy structural confirmation by 13C NMR spectral data and biological activities. Tetrahedron 2003;14 325-331.

[33] Sakamoto S, Hatakeyama M, Ito T, Handa H. Tools and methodologies capable of isolating and identifying a target molecule for a bioactive compound. Bioorganic and Medicinal Chemistry 2012; 20 1990-2001.

[34] Dragull K, Beck JJ. Isolation of Natural Products by Ion Exchange Methods. Methods in Molecular Biology 2012; 864 189-219.

[35] Dufresne C. Isolation by Ion Exchange Methods. In Natural Products Isolation Connell RJP. New Jersey: Humana Press; 1998.

Chapter 3

Affinity Chromatography and Importance in Drug Discovery

Özlem Bahadir Acikara, Gülçin Saltan Çitoğlu, Serkan Özbilgin and Burçin Ergene

Additional information is available at the end of the chapter

http://dx.doi.org/10.5772/55781

1. Introduction

Affinity chromatography which is known as a liquid chromatographic technique for separation and analysis of biomolecules based on their biological functions or individual structures has become increasingly important and useful separation method in pharmaceutical science, biochemistry, biotechnology and environmental science in recent years [1]. This technique is especially known as the most specific and effective technique for protein purification [2]. Separation of the biomolecules is based on highly specific biological interactions between two molecules, such as enzyme and substrate. These interactions, which are typically reversible, are used for purification by placing one of the interacting molecules, referred to as affinity ligand, onto a solid matrix to create a stationary phase while the target molecule is in the mobile phase [3]. Any component can be used as a ligand to purify its respective binding partner. Some typical biological interactions, frequently used in affinity chromatography, can be given as;

- Enzyme. ↔ substrate analogue, inhibitor, cofactor.
- Antibody ↔ . antigen, virus, cell.
- Lectin ↔ . polysaccharide, glycoprotein, cell surface receptor, cell.
- Nucleic acid ↔ . complementary base sequence, histones, nucleic acid polymerase, nucleic acid binding protein.
- Hormone, vitamin ↔ . receptor, carrier protein.
- Glutathione ↔ . glutathione-S-transferase or GST fusion proteins.

- Metal ions ↔ . Poly (His) fusion proteins, native proteins with histidine, cysteine and/or tryptophan residues on their surface [4-5].

In case a ligand is immobilized on a polymeric carrier, usually by covalent coupling, and filled in a column, it is possible to separate the substances which have affinity to the ligand and the other substances. As the solution containing the biologically active substance applied to the column, the compounds which have no affinity to the insoluble ligand will pass through the column and the biologically active compound will be captured on the column, in favorable conditions. The sorbed compounds can then easily be dissociated by changing the external conditions, such as ionic strength, pH, solvent, temperature etc. or alternatively by using dissociating agents [6-7]. As a result, it is possible to isolate and purify the analyte or make quantitative analysis with a suitable, immobilized ligand by means of molecular recognition [1-2].

Macromolecules such as proteins, polysaccharides, nucleic acids differ only in their physico-chemical properties within the individual groups and their isolation on the basis of these differences is therefore difficult and time consuming. Considerable decreases may occur during their isolation procedure due to denaturation, cleavage, enzymatic hydrolysis, etc. The ability to bind other molecules reversibly is one of the most important properties of these molecules. The formation of specific and reversible complexes of biological macromolecules can serve as basis of their separation, purification and analysis by the affinity chromatography [6].

Affinity chromatography is one of the oldest forms of liquid chromatography method [8]. The first use of the idea of affinity chromatography may be considered as the isolation of α-amylase by using an insoluble substrate, starch, in 1910 by Starkenstein [6,9]. Similar studies with starch and amylase were carried out in the 1920s through 1940s by other investigators. In another study polygalacturonase was used as a support and ligand for the adsorption of alginic acid, the purification of pepsin through the use of edestin, a crystalline protein and the isolation of porcine elastase with powdered elastin were also performed. Afterwards Willstatter et al. enriched lipase by selective adsorption onto powdered stearic aicd [10]. The majority of the previous studies related purification of the enzymes. However the selective purification of antibodies with biological ligands was also being conducted. In 1920, it was reported that antibodies can recognize and bind substances with a specific structure, "antigens" [8]. This principle is firstly used in order to isolate rabbit anti-bovine serum albumin antibodies on a specific immunoadsorbent column consisting bovine serum albumin coupled to diazotized *p-aminobenzyl*-cellulose [10]. According to this approach, antibodies were isolated using urease and exhibited that these antibodies were proteins [8].

Separation procedure in affinity chromatography can be simply illustrated as shown in Figure 1. A sample containing the compound of interest is applied to the affinity column in the presence of mobile phase which was prepared in suitable pH, ionic strength and solvent composition for solute-ligand binding. This solvent which is referred as the application buffer presents the weak mobile phase of an affinity chromatography. While the sample is passing through the column compounds which are complementary to the affinity ligand will bind. However other solutes in the sample will tend to be washed off or eluted from the column as

http://dx.doi.org/10.5772/55781

nonretained compounds. After all nonretained components are washed off the column, binding solute or together with ligand as solute-ligand complex are eluted by applying a solvent. This solvent which is referred as elution buffer represents the strong mobile phase for the column. Later all the interested solutes are eluted from the column, then application buffer is applied and the column is allowed to regenerate prior to the next sample application [4,8].

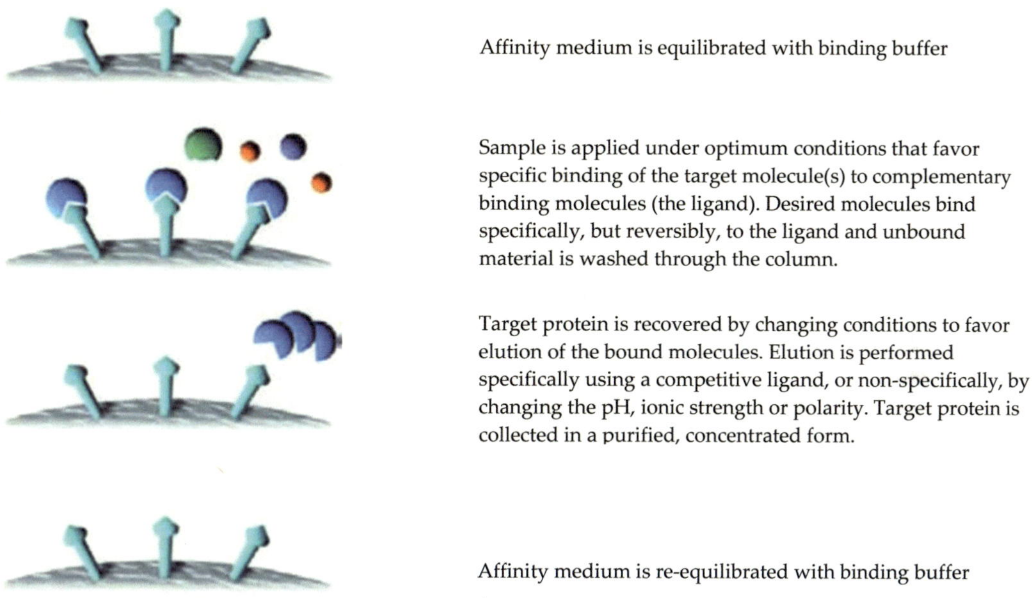

Figure 1. Separation procedure in affinity chromatography

The conditions in which the sample is applied to the column are chosen considering the conditions which the interaction between analyte and ligand is strong, mostly resembling the natural conditions of the analyte and ligand. The content apart from the analyte passes through the column without or with weak binding to the ligand while the analyte is retarded. After the analyte is obtained generally by using an elution buffer, the column is regenerated by washing with the application buffer in order to prepare the column for the next injection [1]. In the Figure 2, a typical scheme of an affinity chromatography application is shown.

As it is defined above; this technique is based on the interactions between specific bioactive substances, so the ligands are supposed to be originally biological substances, nevertheless columns with nonbiological ligands are also available and the same term "affinity chromatography" is used for the techniques performed by using these ligands. In order to distinguish the techniques according to the origin of the ligand, affinity chromatography with biological ligands may be termed as "bioaffinity chromatography" or "biospesific adsorbtion" [1]. The wide application potential of affinity chromatography leaded to the development of derived techniques some of which are listed below [7].

- Immunoaffinity chromatography

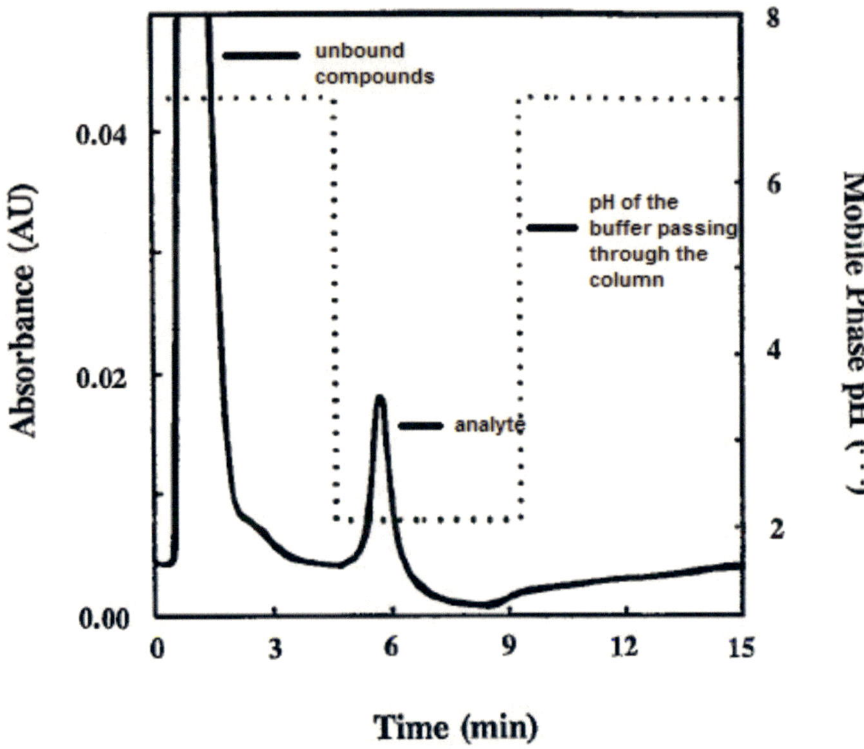

Figure 2. An example of a typical scheme of an affinity chromatography application [1].

- High performance affinity chromatography
- Affinity density perturbation
- Library-derived affinity ligands
- Lectin affinity chromatography
- Affinity partitioning
- Dye-ligand affinity chromatography
- Affinity electrophoresis
- Affinity capillay electrophoresis
- Centrifuged affinity chromatography
- Filter affinity transfer chromatography
- Affinity precipitation

- Avidin-biotin immobilized system
- Affinity tails chromatography
- Affinity repulsion chromatography
- Perfusion affinity chromatography
- Theophilic chromatography
- Weak affinity chromatography
- Receptor affinity chromatography
- Membrane-based affinity chromatography
- Molecular imprinting affinity
- Metal-chelate affinity chromatography
- Covalent affinity chromatography
- Hydrophobic chromatography

Affinity chromatography utilizes specific and irreversible biological interactions between a ligand covalently coupled to a support material and its complementary target. The solid support and ligand covalently attached on it, selectively adsorbs the complementary substance from the sample. The unbound part of the sample is removed and the purified substance can easily be recovered [11]. Selectivity of the ligand, recovery process, throughput, reproducibility, stability and economical criteria are some of the factors that influence the success of affinity chromatography process [9]. Successful affinity purification requires a certain degree of knowledge and understanding of the nature of interactions between the target molecule and the ligand to help determine the selection of an appropriate affinity ligand and purification procedure [3]. Therefore prior to start the process, materials and specifications listed below need to be selected [11]:

- Support material
- Activation method
- Ligand
- Immobilization method
- Conditions for adsorption and desorption

1.1. Support material

For successful separation in affinity chromatography, the important parameter is that solute of interest should be bound firmly and specifically while leaving all other molecules. This requires that the support within the column contain an affinity ligand that is capable of forming a suitably strong complex with the solute of interest [8]. The other important property is that the, support material must be biologically and chemically inert to avoid the unspecific bindings [8,11] which requires that the support has a chemical character that is very similar to that of the medium in which it is operating. Since almost all affinity separations occur in aqueous solutions, the support should thus be as hydrophilic as possible. As a rule, the mobile phase used in affinity separations has a low ionic strength. The support should therefore contain as few charges as possible to prevent ionic interactions. Many supports are available which have desired properties or they gain such characteristics by hydrophilic coating [8]. Generally solid materials are used as support material though some soluble macromolecular materials are sometimes preferred for two-phase aqueous affinity partition processes. Uniformity in particle size and ease of the activation process are also required for support material that is used in affinity chromatography applications [11]. For the affinity chromatography at low pressure, nonrigid gels with large particle size are generally used as support materials while materials with small, rigid particles or synthetic polymers which are stable under high pressure and flow rates are used in high performance affinity applications [8,12].

There are many commercially available support materials for affinity chromatography can be divided into three groups as; natural (agarose, dextrose, cellulose); synthetic (acrylamide, polystyrene, polymethylacrylate) and inorganic (silica, glass) materials [7,13]. The most popular support material is agarose [13]. Agarose was used in the first modern application of affinity chromatography and still the most commonly preferred one [8]. Agarose consists of alternatively linked 1,3 bound β-D- galactopyranose and 1,4 bound 3,6-anhydro-*α*-*L*-galactopyranose, as shown in Figure 3 [8,14]. Agarose gels are stable to eluants with high concentrations of salt, urea, guanidine hydrochloride, detergents or water-miscible organic solvents but its stability is less beyond pH 4-9. To increase the thermal and chemical stability, cross-linked agarose is prepared. Cross-linked agarose is commercially available (Sepharose) and it can be used with many solvents, over pH 3-14 and at high temperatures up to 70°C [11]. However, strong acids, oxidizers as well as some rare enzymes may be harmful to agarose due to their damaging effects. On the other hand mild acid hydrolysis increases the quantity of sterically available galactose residues and turns agarose into an excellent sorbent for galactose binding proteins [14]. Due to its large beads and macroporous, accessible pore structures, agarose is well designed for use with large molecules. High capacity, presence of functional groups, good chemical stability especially at high pH, low non-spesific binding and good reproducibility are the advantage of agarose. Some properties of agarose such as solubility in hot water and non-aqueous solutions, sensitivity to microbial degredation and lack of rigidity restrict the usage under low or medium pressure [15]. Furthermore agarose must not be frozen or air dried. It is sold under several trade names, including Sepharose Fast Flow or Affi-Gel.

Cellulose is another example of polysaccharides which is used as support in affinity chromatography. Cellulose has a historical significance. Phospo- and DNA-cellulose are used especially for DNA related separations [14]. Antibody and enzyme purifications have also been carried out. However its fibrous and non-uniform character limits its use since cellulose detains macromolecules [11].

Dextran which is a linear *α*-1,6-linked glucose polymer produced by *Leuconostoc mesenteroides* is also used in affinity chromatography. Sephadex (cross-linked by glyceryl bridges) and

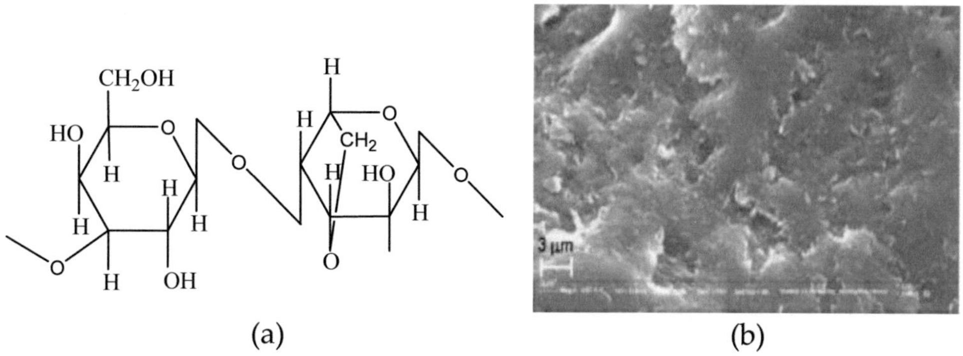

Figure 3. (a). Structure of agarose **(b).** SEM image of agarose [16]

Sephacryl (allyldextran cross-linked by *N,*N′-methylenebisacrylamide) are the two types of dextran gel used in separation. Sephadex is mainly used as a glucose polymer and employed for purification of many molecules such as lectins from *Helix pomatia* and *Vicia faba*, exoamylase from *Pseudomonas stutzeri* [14].

Polystyrene (Figure 4), which is a polymeric support is also unsuitable in its original form for affinity separations due to the highly hydrophobic character. Native polystyrene, which is often used as a reversed-phase material, must be first rendered hydrophilic by one of various surface-coating techniques before used in other chromatographic methods [8].

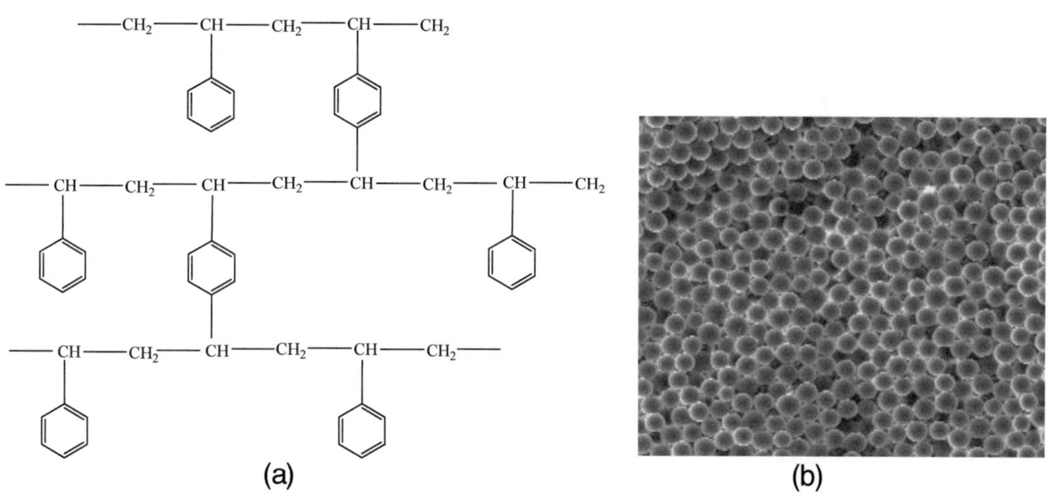

Figure 4. (a) Structure of polystyrene divinilybenzene **(b)** SEM image of polystyrene beads(x500) [17]

Polymeric supports based on polyacrylamide are synthesized by copolymerization of acrylamide and a cross-linking reagent and can be used directly in affinity chromatography due to its more hydrophilic properties than polystyrene supports. Polyacrylamide gels are either soft or have small pores. However this gel is resistant against enzymatic attacks and does not absorb biomolecules, thus it is used widely despite its mechanic disadvantages [8,14].

Inorganic materials such as porous glass and silica are used when the extreme rigidity of the support material is needed [14]. Silica is especially used in the high-performance liquid affinity chromatography (HPLAC) or high-performance affinity chromatography (HPAC). Silica-based materials (Figure 5) are basically hydrophilic and they are suitable for affinity chromatography after they are modified at their surface. The native surface of silica is primarily covered with silanol groups which are weak acids and give strong negative charge to silica's surface at neutral pH. Irreversible adsorption of solutes (protein) can occur due to these charges and in combination with other binding forces. However, several methods can be used to render this surface inert toward such solutes, including polymer coating techniques and reactions between silica and alcohols or trialkoxysilanes [8]. They are also soluble at pH above 8 thus pH is an important parameter that limits the usage of silica [14].

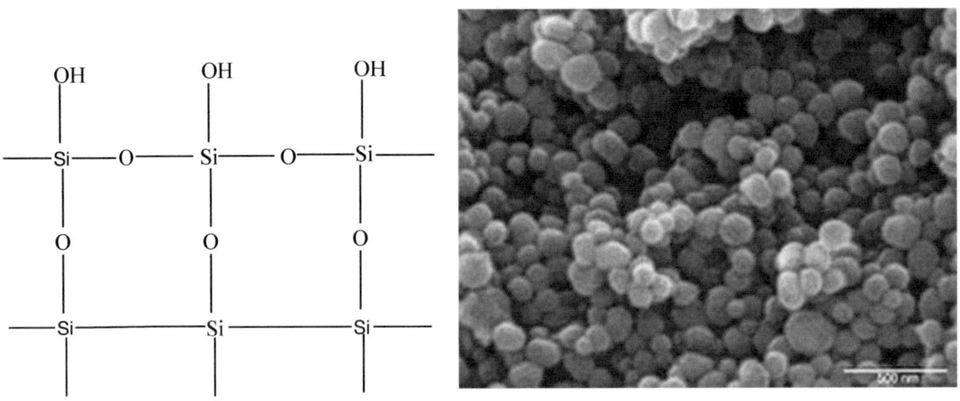

(a) (b)

Figure 5. (a) Structure of silica **(b)** SEM picture of typical silicagel [18]

A support material should be inert toward solutes. On the other hand easy coupling with ligand is also desired. Support materials are rich in hydroxyl groups, therefore attachment of ligands have been focused mainly on using these regions as anchoring points. Ideal affinity support should allow unhindered access of a solute to the immobilized ligand. For a macromolecular solute, this requires a support that has large pores. Renkin equation can explain these pore sizes to be that allows one to estimate the effective diffusion coefficient (D_{eff}) of a solute in a porous material.

$$D_{eff} = D\ K_D \varepsilon_p [1 - 2.10(R_s / R_p) + 2.09(R_s / R_p)^3 - 0.095(R_s / R_p)^5] / t$$

In this equation, R_s/R_p is the ratio of the solute's radius (R_s) to the pore radius (R_p), ε_p is the particle porosity, τ is the tortuosity factor, K_D is the distribution coefficient for the solute, and D is the diffusion coefficient for the solute in free solution. By inserting different values for the

http://dx.doi.org/10.5772/55781

ratio R_s/R_p, one finds that the pore diameter should be at least five times the diameter of the solute to avoid severely restricted rates of diffusion.

For a protein in normal size (i.e., a diameter around 60 Å), a ratio of five for R_p/R_s means that the support pores should be in the range of 300 Å. Several common supports are available with such pore sizes. Support materials with very large pores give essentially unhindered diffusion for most solutes, but they also have a smaller surface area per milliliter of bed volume than supports with smaller pores. This reduced surface area leads to a diminished binding capacity. As a rule, a pore size of 300 to 700 Å is usually a good compromise in most situations encountered in affinity chromatography, since this gives fairly unrestricted diffusion for most biomolecules while also providing a relatively large surface area for retention [8].

Particle diameters of the affinity supports are available in a wide variety. These range from HPLC-type materials with diameters of 10 μm or less to large particles for preparative work that have diameters of 400 μm. The purpose of the separation, mechanical properties of the support and the characteristics of the sample are important factors on the selection of particle size of the support. From a theoretical viewpoint, it is always advantageous to have a small particle size, since this will promote fast mass transfer of a solute between the outer flow stream and interior of a support particle. Sample molecules are transported down through the column by the flow of the mobile phase in the spaces between the support particles. To reach the affinity ligands, these molecules must diffuse through the stagnant mobile-phase layer surrounding the particles (i.e., the film model) and proceed to the inside pore network (Figure 6). It is where the sample molecules will finally bind to the affinity ligand. When the retained molecules are eluted, the same steps occur but in a reversed order. Smaller support particles mean shorter diffusion distances, since they have shorter pores and a thinner stagnant mobile phase layer around and in the support. This results in shorter times needed for diffusion.

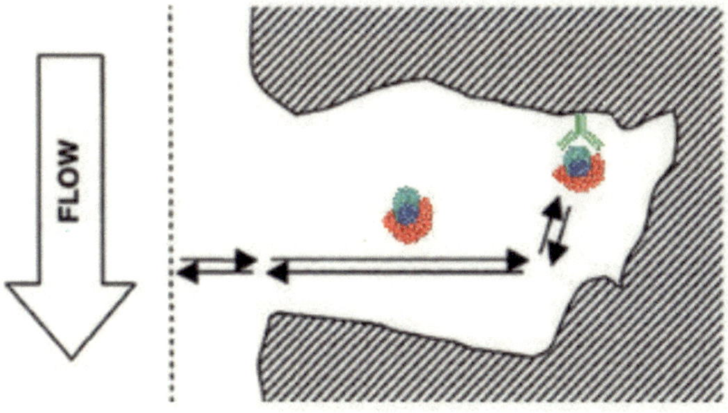

Figure 6. Transport processes that occur in a chromatographic column.

In preparative affinity chromatography, relatively large support particles are often used, making intraparticle diffusion the main factor limiting efficiency. In this case, diminishing the particle size will increase the rate of movement of solutes between the support and surrounding flow stream, giving an improved column performance. It is this effect that was the original driving force behind the use of smaller supports in affinity columns, thus giving rise to the technique of HPLAC. Under such conditions, a decrease in particle size by a factor of five can make it possible to increase the flow rate by up to 25-fold and still retain good chromatographic performance. This results in a dramatic improvement in the productivity of the system. However, a point is eventually reached when a decrease in particle diameter no longer gives a proportional improvement in an affinity column's performance. This has been observed in many analytical-scale systems that use HPLC-type supports with particle sizes less than 10 μm in diameter. Under these conditions, diffusion in the particle is now relatively fast, and it is the adsorption/desorption of sample molecules to and from the affinity ligand that becomes the limiting factor in speed and efficiency. Although better efficiency is always obtained with small support particles, using a small particle size tengenders some difficulties. One problem is the much higher flow resistance of these smaller particles. This increased flow resistance may lead to bed collapse when using soft gels such as agarose. And, although supports like silica can tolerate the higher pressures, that results, these will require the use of more expensive pumps to work at such pressure, as is generally done in HPLC. Another route that could be taken with small affinity supports is to use a short and wide column instead of a long and narrow one. The advantages of this are that the shorter, wider column can be run at higher flow rates without creating high-pressure drops. Another drawback with small particle sizes, especially in preparative work, is the increased danger of fouling that exists when particulate contaminants are in the feed stream or sample. This occurs because the interstitial spaces in a bed of small particles can be too narrow for such agents to pass through. Such fouling will increase the flow resistance and may lead to bed collapse if the support material does not have sufficient mechanical strength.

As a result of these various requirements, the particle size to pick when designing a new affinity adsorbent will be a compromise between the desired chromatographic performances, properties of the feed stream, and the mechanical strength of the support. Some common selections made in specific cases will be described in the next few sections.

Porous supports like agarose, polymethacrylate, or silica beads are generally used in current applications of affinity chromatography. However, in the past several years other types of supports have also become available commercially. Many of these newer materials have properties that give them superior performance in certain applications. Materials that fall in this category include; nonporous supports, membranes, flow-through beads, continuous beds and expanded-bed particles.

Nonporous beads with diameters of 1 to 3 μm can be an optimum choice for fast analytical or micropreparative separations, since the limiting factor of pore diffusion is virtually eliminated in these materials. Such beads may also be the best choice for fundamental or quantitative studies of affinity interactions, since the binding and dissociation behavior observed in these materials should be more directly linked with the interactions occurring between solutes and the affinity ligand. However, there is a substantial loss of surface area and binding capacity.

http://dx.doi.org/10.5772/55781

Membranes have been used for affinity chromatography in various formats, such as stacked sheets, in rolled geometries, or as hollow fibers. Materials that are commonly used for these membranes are cellulose, polysulfone, and polyamide. Because of their lack of diffusion pores, the surface area in these materials is as low as it is in nonporous beads. However, the flat geometry and shallow bed depth of membranes keep the pressure drop across them to a minimum degree. This means that high flow rates can be used, which makes these membranes especially well-suited for capturing proteins from dilute feed streams.

As stated earlier, porous supports with a larger diameter facilitate low column backpressures and allow easy passage of contaminants through the column. But it is also necessary to keep the diameter of these supports as small as possible to diminish diffusion distances and thereby improve their chromatographic performance. One solution of these contradictory requirements is to use particles that allow the flow of mobile phase directly through some of the pores. This is done in materials known as *perfusion media* or *through-pore particles*. Flow-through particles were initially developed in the early 1990s for ion-exchange chromatography and were later adapted for use in affinity chromatography.

Flow-through particles have a bimodal pore configuration, in which both small diffusion pores and large flow-through pores are present. Substances applied to a bed of this support are transported by mobile phase flow to the interior of each particle, leaving only short distances to be covered by diffusion to the support's surface (Figure 7). This combination leads to a dramatic improvement in performance compared with standard porous particles of the same size. This improvement is most pronounced in situations where slow diffusion is a limiting factor, such as in the chromatography of large molecules (e.g., proteins) at high flow rates.

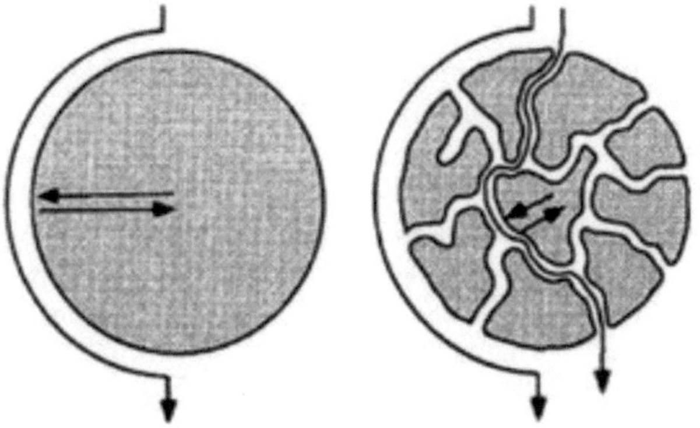

Figure 7. Comparison of a particle with normal porosity versus a particle that contains flow-through pores. The normal particle has long diffusion distances, whereas the flow-through particle has short diffusion distances.

Another format developed in 1990s was the *continuous bed* or *monolithic support*. Continuous bed supports consist of a single piece of material intersected by pores large enough to support chromatographic flow through the bed. Continuous beds have been developed using many well-known chromatographic materials, such as polyacrylamide, silica, polystyrene/polymethacrylates, cellulose, and agarose. Most of these continuous beds have two types of pores: large flow-carrying pores and smaller diffusion pores.

The preparation of a continuous bed is usually straightforward. These beds can often be prepared directly in a chromatographic column, thereby avoiding the time-consuming steps of size classification and column packing that are normally needed with particle-based supports. Reports using continuous beds in affinity chromatography have shown that the efficiency of these materials is as good as that for particle-based supports.

In order to avoid column clogging, various pretreatment methods like filtration and centrifugation are often necessary to remove particulate matter from samples. To cut down on the need for such methods, a new class of adsorbents has recently been developed to handle viscous and particle-containing feed streams. These materials are known as *expanded-bed adsorbents*. In expanded-bed chromatography the direction of mobile phase flow is upward through the column and is fast enough to fluidize the support particles in the column. This causes the column bed to expand. This expansion makes the interstitial spaces in the column bed larger so that solid contaminants like cells and cell debris can pass through, thereby avoiding column clogging.

Another new type of expanded-bed adsorbent uses a thin layer of active material (i.e., derivatized agarose) that surrounds a heavy core. These adsorbents have small diffusion distances for biomolecules along with a higher density than other expanded-bed particles. The advantage of this combination is that it allows better chromatographic efficiencies to be obtained at higher flow rates [8].

1.2. Ligand

Ligands are the molecules that bind reversibly to a specific molecule or group of molecules, enabling purification by affinity chromatography [4]. These molecules which play a major role in the specificity and stability of the system are essential for affinity chromatography [13]. The selected ligand must be capable of selectively and reversibly binding to the substance to be isolated and have some groups which are available for modifications in order to be attached to the support material. It is very important to ensure that the modifications do not reduce the specific binding affinity of the ligands. There are general ligands such as dyes, amino acids, Protein A and G, lectin, coenzyme, methal chelates as well as specific ligands such as enzymes and substrates, antibodies and antigens [19].

Affinity ligands are classified as synthetic and biological. Biological ligands are obtained from natural sources such as RNA and DNA fragments, nucleotides, coenzymes, vitamins, lectins, antibodies, binding or receptor proteins, or in vitro from biological and genetic packages, employing display techniques including oligonucleotides, peptides, protein domains and proteins. Synthetic affinity ligands are generated either by de novo synthesis or modification of existing molecular structures (triaznyl nucleotide-mimetics, purine and pirimidine derivatives, non-natural peptides, triazinyl dyes, other triazine-based ligands, oligosaccharide and boronic acid analogues). These can be generated by rational design or selected from ligand libraries. Synthetic ligands are generated using three methods;

- The rational method features the functional approach and structural template approach.
- The combinatorial method relies on the selection of ligands from a library of synthetic ligands synthesized randomly.
- The combined method employs both methods the ligand is selected from an intentionally prepared library based on a rationally designed ligand.

Many parameters have to be taken into account in order to select appropriate ligand. Table 1 exhibits the advantages and disadvantages of synthetic and biological ligands. Selectivity and affinity are the main advantages of biological ligands. Such ligands can be generated by *in vitro* evolution approaches and selecting from large combinatorial ligand libraries based on biological/genetic packages. Protein ligands display special advantages for example; higher affinities, higher proteolytic stability, preservation of their biological activity when produced by fusion to a different protein or domain. However these ligands can be expensive and unstable to the sterilization and cleanin conditions used in manufacturing process of biologics because of their biological origin, chemical nature and production methods. There is high contamination risk of the end-product with potentially dangerous leaches, in addition to possible contaminants originated from the biological source [20].

	Synthetic ligands	**Biological ligands**
Capacity	High	Low to medium
Cost	Low to medium	Medium to high
Selectivity	Medium to high	Very high
Stability	High	Low to medium
Toxicity	Medium	Low

Table 1. Comparison of biological and synthetic ligands [20]

Despite the advantages of the affinity chromatography technique, its use is limited due to high cost of affinity ligands and their biological and chemical instability. The development of methods for production of stable synthetic ligands has enabled "utilization of these materials in large scale. For the design of synthetic ligands, information about structure of the target protein and a potential binding site are required, thus a structure-based design can be achieved, in case correct prediction of the ligand's comformation and the binding affinity of the designed ligand. Function-based design can be applied when the structure of the target is not known [9]. Substantially, selection and design of ligands may be performed by using a template which is a part of a natural protein-ligand couple, modelling a molecule which complements the binding sites of the target or directly resembling the natural interactions [2].

High selectivity of the biological ligands is a benefit; however these ligands have some handicaps, such as their low binding capacity, cost-efficacy issues, some problems in scale-up and purification process. Hence, synthetic ligands may offer a solution for these issues and enable to provide selectivity, efficacy and inexpensiveness in a body. Biomimetic textile dyes which are developed in 1970s are the most known synthetic ligands. The use of these dyes in biopharmaceutical field is limited due to some issues such as selectivity, purity and toxicity. These complications have led to new researches and developments about biomimetic dyes and new ligand design techniques [2].

The selection of the ligand may be done according to the specific binding site of a target, but this manner of selection may fail owing to the fact that immobilization process may change binding affinity. It is known that the affinity of the target to the ligand is dependent on the features of the target as well as support material, activation and coupling chemistry. Some other techniques other than using free ligand solution in order to predict the conditions of three-dimensional matrix. On the purpose of ligand selection, a great number of alternatives may be tested for binding the target or work with more accurate options by employing ligand design techniques. Therefore the idea to combine chemistry with computational tools has accelerated the developments on this field [2]. Along the development of affinity chromatography techniques, different laboratories are established with the purpose of collection of several ligands for affinity chromatography [9].

Protein-structure-based design of the ligands depends on the correct prediction of the structure of the target protein and the binding site. Apart from this, protein-function-based design is applicable in case the conformation of the target protein is not known. This method is based on the integration of some known properties of the ligand such as an essential molecular structure, a functional group or a derivative of some parts of the structure [9]. The design of a ligand requires several steps to fulfil [2]:

1. Determination of the binding site or possible biological interactions to use as a template for the modelling,
2. Initial design of the ligand using this template,
3. Preparation of a ligand library and chromatographic evaluation,
4. Selection of the ligand of interest,
5. Optimisation and chromatographic evaluotion of the adsorbent following the immobilization.

Beyond these design methods, some combinatorial approaches have been developed on the purpose of ligand selection. Synthetic peptide libraries which include all sequences for a length of a protein structure are one of these approaches. By means of these libraries, *in vitro* prediction of the action of the library mixture as it passes through the surface where the protein of concern is immobilized is possible. The ligands which possess affinity to the immobilized protein are

http://dx.doi.org/10.5772/55781

suggested as ligands for the affinity chromatography application. It is also of choice to use phage libraries and a screening method known as biopanning. Phage display method allows determining suitable ligands not only for peptides and proteins but also for nonpeptide structures (Figure 8).

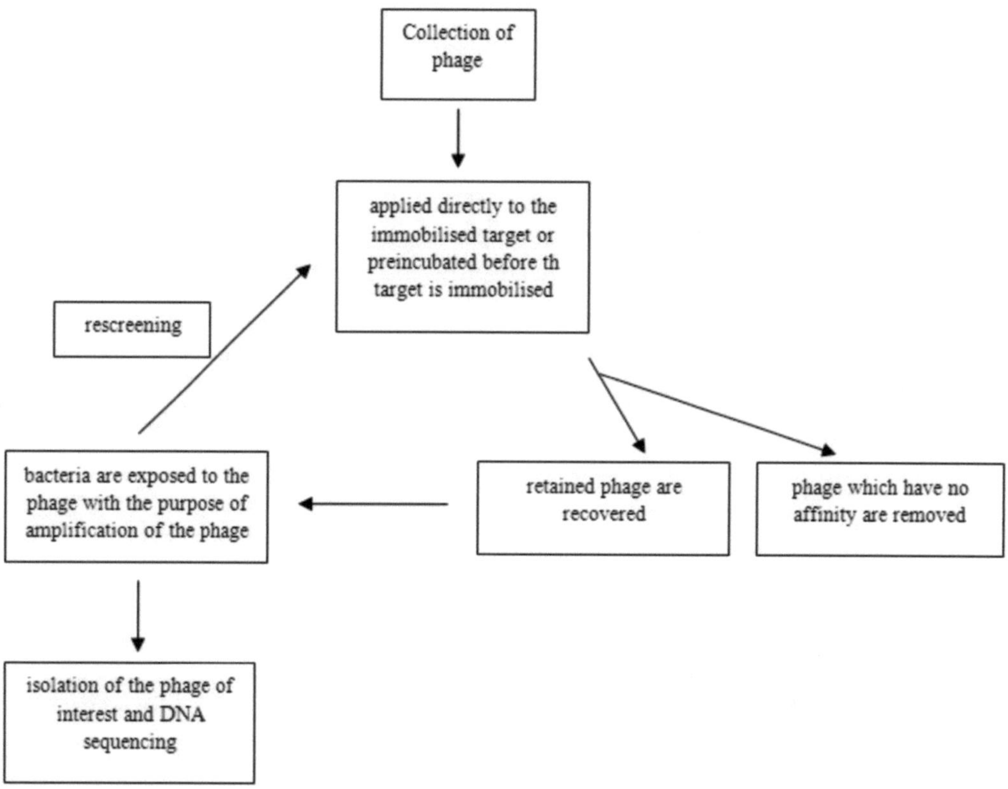

Figure 8. Selection of the phage from the phage-displayed combinatorial peptide library [9]

A collection of phage is incubated with the target and retained ligands are considered to be candidates for ligand [9]. At the end of the cycle, the process is repeated in order to increase the amount of the protein which has required binding features. The sequence of the protein is provided from the viral DNA. Phage display libraries are indisputably useful especially for epitope mapping, vaccine development, bioactive peptides and some non-peptide structures and the ligands determined using this method are appropriate for chromatographic analyses, nevertheless there are some cases that limit the use of this technique, such as some optimization problems and some issues as working in large scale as well as the limitation of the application since peptides does not work unless it is a part of the phage, not in free solution [2].

Although its use in ligand selection for large scale of affinity chromatography is not wide, ribosome display and systematic evolution of ligands by exponential enrichment (SELEX) may be mentioned as another approach and a potential ligand design and selection method due to its versatility and rapidity [9]. Ribosome display method enables to select and develop a protein library *in vitro* [2]. The principle of ribosome display is depicted in Figure 9. A collection of DNA encoding the selected peptide is exposed to an *in vitro* transcription and translation process, then in favourable condition, the complex of mRNA, peptide and ribosome since the stop codon does not exist. Thereafter the complex is passed thtough the immobilised target and the peptides which possess high affinity to the target are retained. At the end of this process, it is possible to seperate mRNA usually by EDTA, then by means of reverse transcription, DNA are attained and amplified [9].

SELEX is a widely used technique for screening of aptamers which are nucleic acid ligands. According to this method, a pool of DNA with a random sequence region attached to a constant chain is constituted by amplification then transcribed to RNA. RNA pool is separated according to the affinity of RNA molecules to a target protein. DNA molecules obtained by reverse transcription from retarded RNA molecules are amplified and the cycle is repeated.

The selection of the ligand may be designed according to the structure of the target protein as well. Under the favor of combination of the structure-based design and combinatorial chemistry, the efforts to synthesis a convenient structure are minimized [9].

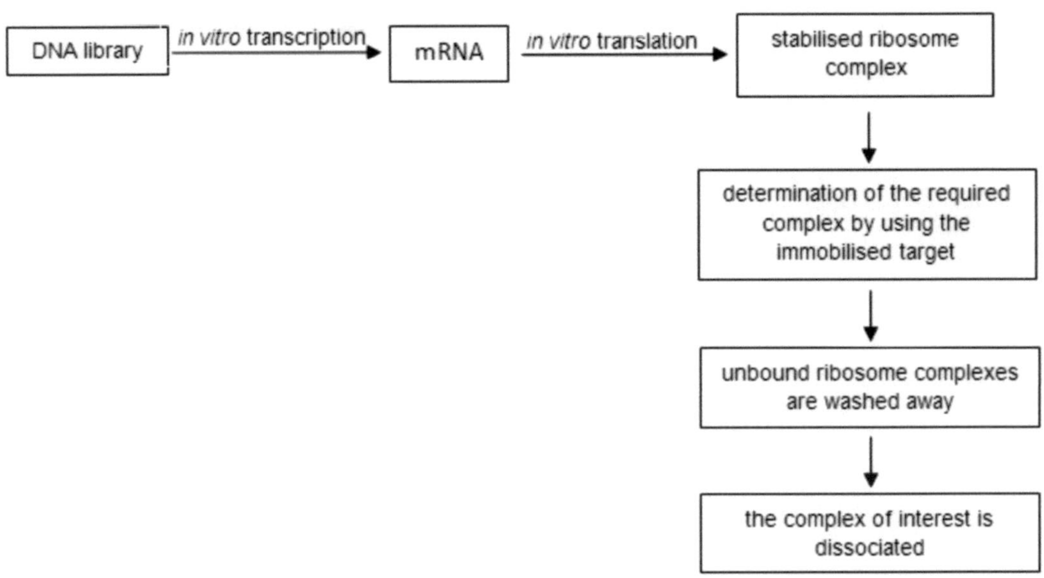

Figure 9. Ribosome display method [9]

1.3. Immobilization

The immobilized ligand is an essential factor that determines the success of an affinity chromatographic method [12]. The method which is used for affinity ligand immobilization is important because actual or apparent activity of the final column can be affected. Decrease in ligand activity which result in multisite attachment, inappropriate orientation or steric hindrance can be observed if the correct procedure is not pursued [21]. Several methods are available to couple a ligand to a pre-activated matrix. The correct selection of coupling method depends on the ligand characteristics [4].

Before ligands are coupled matrix is activated. Among several methods used for activation, the cyanogen bromide activation is the most frequently preferred. Activation using this method produces a highly reactive cyanate ester [7]. The ligands are attached to the support via primary aromatic or aliphatic amino groups. High toxicity of cyanogen bromide is the disadvantage of this method [13,19]. Subsequent coupling of ligands to the activated matrix results in an isourea linkage. Despite the popularity of this method, the isourea linkage of the ligands causes several problems during the purification procedure, including nonspecific binding due to charge and leakage of the ligand because of instability of the isourea bond. *N*-hydroxysuccinimide (NHS) esters have also been used for immobilizing ligands. The preparation of active esters requires a matrix that contains carboxylic groups. Such matrices can be easily obtained from agarose by activation of the hydroxyl groups with different reagents, including cyanogen bromide, activated carbonates, etc. and successive reaction with ω-amino acids of different sizes depending on the length of the spacer arm required. The NHS ester is then prepared by mixing the carboxylic matrix with dicyclohexylcarbodiimide and NHS. Due to the stability problem a different method based on *N,N,N′,N′*-Tetramethyl (succinimido) uronium tetrafluoroborate can be also used. The covalent attachment of ligands to such activated carriers provides the production of stable amide bonds. Another method for activating polysaccharides is the use of *N′N*-disuccinimidyl carbonate (DSC), which forms highly reactive carbonate derivatives with polymers containing hydroxyl groups. These derivatives react with nucleophiles under mild, physiological conditions (pH 7.4), and the procedure results in a stable carbamate linkage of the ligand coupled to the carrier. The immobilization of different ligands (e.g., enzymes, enzyme inhibitors, antigens and antibodies) on activated carbonate carriers has been achieved, together with excellent maintenance of biological activity of the proteins [7]. Pre-activated commercial matrices are also available (Table 2) to avoid many steps and problems of chemical activation process. A wide range of coupling chemistries, involving primary amines, sulfhydryls, aldehydes, hydroxyls and carboxylic acids are available for covalently attaching ligands to the matrices [12]. The use of commercially available, pre-activated media is recommended to save time and avoid the use of the potentially hazardous reagents that are required in some cases [4].

After the activation of the support material, it is ready for the immobilization process of the ligand. In case the ligand is a small molecule, steric hindrance will occur between the immobilized support and the compound of interest (Figure 10). This may reduce or totally block specific binding of the substance. Use of the supports having a spacer arm attached or attachment of a spacer molecule to the support before immobilization of the ligand generally solves this problem. Spacer arm keeps ligand at a suitable distance from the surface of the support (Figure 9), thus the substance of interest will not be prevented to attach to the immobilized ligand. It is possible to bind spacer arms directly to the support prior to the imobilization of the ligand. Then a secondary reaction provides the attachment of ligand to

Product name	Functional group specifity
UltraLink Iodoacetyl resin	-SH
CarboLink Coupling resin	-CHO, C=O
Profinity™ Epoxide resin	$-NH_2$, -OH, -SH
Affi-Gel 10 and 15	$-NH_2$
Pierce CDI-activated resin	$-NH_2$
Epoxy-activated Sepharose™ 6B	$-NH_2$, -OH, -SH
CNBr-activated Sepharose 4 Fast Flow	$-NH_2$
EAH Sepharose™ 4B	-COOH, -CHO
Thiopropyl Sepharose™ 6B	-SH
Tresyl chloride-activated agarose	$-NH_2$, -SH

Table 2. Activated commercially available resins of affinity chromatography

the spacer. The substance of interest doesn't be able to bind the ligand unless the spacer arm is long enough, but it is also possible to shorten the spacer arm in salt buffer [19].

Figure 10. Spacer arm, keeping ligand at a suitable distance from the surface of the support

Properties of an ideal spacer arm are listed below:

1. It should be long enough (at least 3 atoms) to keep the substance at an appropriate distance.

2. It should be inactive not to cause a non-specific binding.

3. It should have bifunctional group for the reaction with both support and the sample [9].

Compounds which have diamine groups such as hexanediamine, propanediamine and ethylenediamine are the most preferred spacer arms used in affinity cromatography. Some other examples of spacer arms are shown in Table 3 [19].

The following step is the immobilization of ligands on the activated matrix by isourea bonds. Immobilization through isourea linkage has some disadvantages including nonspesific binding of the ligand because of the instability of the bonds. Another method for immobilization is to use active esters such as *N*-hydroxy-succimide (NHS) esters. The carboxyl groups required for preparation of active esters can be prepared by activation of hydroxyl groups of

http://dx.doi.org/10.5772/55781

Name	Structure
Alkylamine	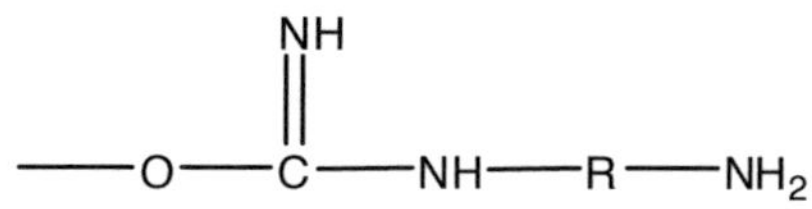
Diamine	$H_2N-R-NH_2$
Polypeptides	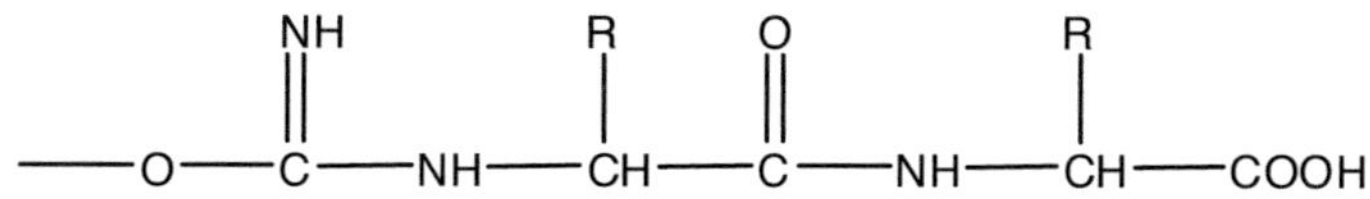
Polyamine	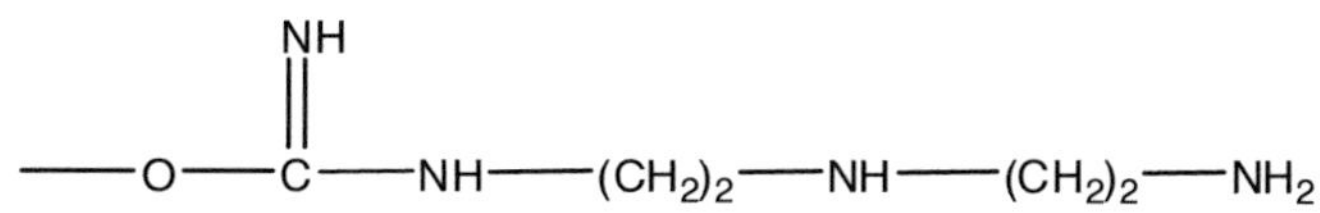
Polyether	$-O(CH_2)_2O(CH_2)_2O(CH_2)_2OH$
Amino acid	NH_2RCOOH

Table 3. Some examples for spacer arms and their structures

agarose. The ligands attach to this type of matrix via amide bonds. It is also possible to activate polysacharides by formation of highly reactive carbonate derivatives. In this case the polymer which contains hydroxyl groups is activated by the use of *N'N*-disuccinimidyl carbonate (DSC). The resultant carbonate derivatives create stable carbamate bonds with nucleophiles under mild, physiological conditions. Immobilization methods can be categorized as follow (Figure 11) [7].

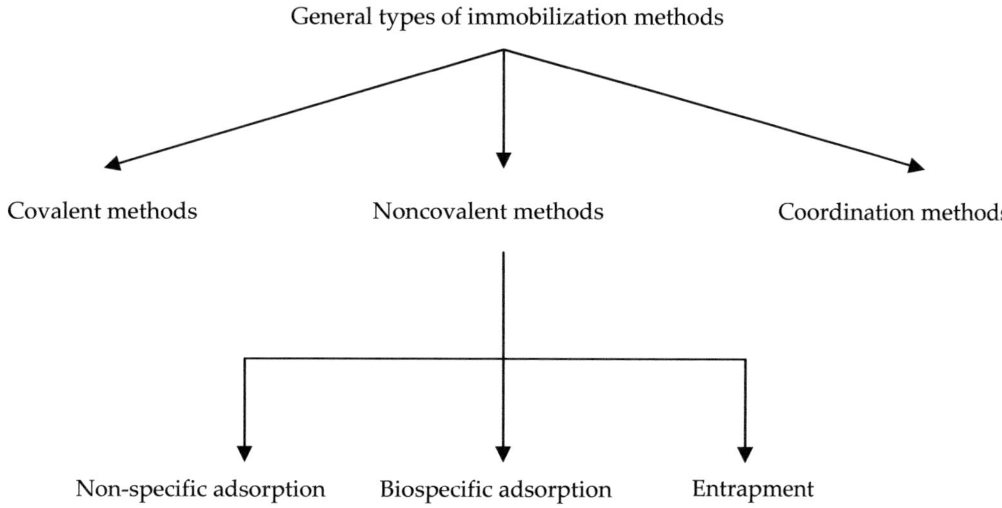

Figure 11. Immobilization methods used in affinity chromatography

1.3.1. Noncovalent immobilization technique

The simple adsorption of ligand to surface, binding to a secondary ligand, or ligand immobilization through a coordination complex can be classified as this type of immobilization. This technique can be subdivided as follow;

a. Nonspecific Adsorption; It is based on the attachment of ligand to support that has not been specifically functionalized for covalent attachment. Adsorption of the ligand to a support depends on the chemical characteristics of both the ligand and support. Columbic interactions, hydrogen bonding, and hydrophobic interactions involve in this type of immobilization.

b. Biospecific Adsorption; In this type of noncovalent immobilization method the ligand of interest bind to a secondary ligand attached to the support. Although a variety of secondary ligands can be used for this purpose, two of the most common are avidin and streptavidin for the adsorption of biotin-containing compounds and protein A or protein G for the adsorption of antibodies.

c. Coordination Complexes; A coordination complex can be used to prepare an immobilized ligand in some cases. This is used to place metal ions into columns for immobilized metal-ion affinity chromatography (IMAC) which is based on the formation of a complex between a metal ion and electron donor groups.

1.3.2. Covalent immobilization methods

Covalent immobilization is the most popular method in affinity chromatography. In this method, it is necessary to activate the ligand and/or the support first. Activation of the ligand can be conducted when it is desired to couple this ligand through a specific region. An example is the creation of aldehydes in the carbohydrate regions of an antibody for its attachment to a support that contains amines or hydrazide groups. The use of an activated support is more common for ligand immobilization but tends to be less specific in nature. Examples include the immobilization of proteins through their amine groups to supports activated with *N*-hydroxysuccinimide or carbonyldiimidazole. The support used for covalent immobilization must meet several requirements. First, sufficient number of groups for activation and ligand attachment should be. Hydroxyl groups on the support are employed in most covalent coupling methods. Depending on how its surface is activated, a support can be used to immobilize ligands through their amine, sulfhydryl, hydroxyl, or carbonyl groups, among others.

1. Amine-Reactive Methods; Amine groups is often used for the immobilization of proteins and peptides. Specific methods are cyanogen bromide method, reductive amination, *N*-hydroxysuccinimide technique, and carbonyldiimidazole method.

a. Cyanogen Bromide Method ; The cyanogen bromide (CNBr) method was the first technique used on a large scale for immobilizing amine-containing ligands and involves the derivatization of hydroxyl groups on the surface of a support to form an active cyanate ester or an imidocarbonate group. Both of these active groups can couple ligands through primary amines, but the cyanate ester is more reactive than the imidocarbonate. The CNBr method utilizes relatively mild conditions for ligand attachment, making it suitable for many sensitive biomolecules. But one problem with this approach is that the isourea linkages obtained by the reaction of CNBr with the support are positively charged at a neutral pH. This means that these groups can act as anion exchangers and nonspecific binding can be occur. Other problems with this method include the toxicity of CNBr, requiring the use of adequate safety precautions during the activation process, and the leakage of ligands that can result from CNBr-activated supports (Figure 12).

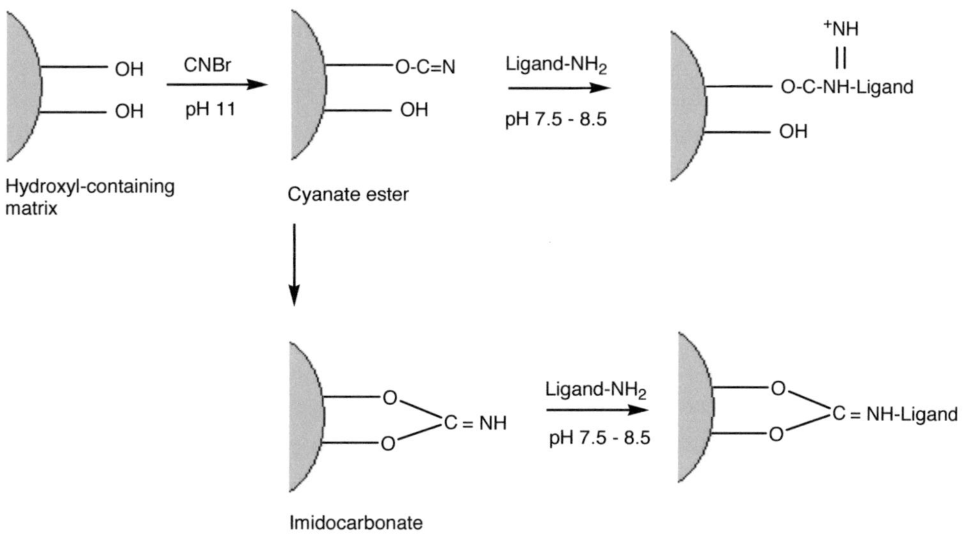

Figure 12. Cyanogen bromide immobilization method patway [8]

b. Reductive amination (also known as the Schiff base method); Reductive amination couples ligands with amine groups to activated periodate is used to oxidize diol groups on the support's surface to give aldehydes. This can be performed directly on carbohydrate-based supports like dextran or cellulose. However, materials like silica or glass must first be treated to place diols on their surface. This can be accomplished by reacting the silica or glass with γ-glycidoxypropyltrimethoxysilane, followed by acid hydrolysis. When an amine-containing ligand reacts with the aldehyde groups, the resulting product is known as a Schiff base. Since this is a reversible reaction, the Schiff base must be converted into a more stable form. This is achieved by including sodium cyanoborohydride in the reaction mixture. Cyanoborohydride is a weak reducing agent that converts the Schiff base into a secondary amine without affecting the aldehydes on the support. After the coupling reaction is completed, the remaining aldehyde groups can be removed by treating the support for a short period of time with a stronger reducing agent (i.e., sodium borohydride) or by reacting these groups with an excess of a small amine-containing agent (e.g., ethanolamine). The Schiff base method is relatively easy to perform and often gives a higher ligand activity than other amine-based coupling methods. This also results in ligands that have stable linkages to the support and that can be used for long periods of time. However, there are some disadvantages of this method. One is the need to work with relatively hazardous agents such as sodium cyanoborohydride and sodium borohydride. Thus, care must be taken to perform this technique with proper ventilation and safety precautions. The use of sodium borohydride for the removal of excess aldehyde groups must also be carried out with caution, since the use of conditions that are too harsh may result in the loss of ligand activity.

c. *N*-Hydroxysuccinimide Method; The N-hydroxysuccinimide (NHS) method is another technique often employed when immobilizing biomolecules through amine groups. This gives rise to the formation of a stable amide bond. There are a number of ways a support can be activated with NHS. The relative ease with which activated supports can be prepared is one advantage of the NHS method. But the fast hydrolysis of NHS esters tends to compete with the immobilization of ligands. This rate of hydrolysis increases with pH and is particularly important when dealing with dilute protein solutions. The half-life of these NHS groups at pH 7 and 0°C is approximately 4 to 5 h and decreases to as little as 10 min at pH 8.6 and 4°C (Figure 13).

d. Carbonyldiimidazole Method; Carbonyldiimidazole (CDI) can also be used to activate supports for the immobilization of amine-containing ligands. This reagent can react with materials that contain hydroxyl groups to produce an acylimidazole, which forms an amide linkage as the result of the interaction with primary amines on a ligand. Supports with hydroxyl groups will react with CDI to produce an imidazolylcarbamate. The reaction of imidazolylcarbamata with primary amines proceeds at pH 8.5 to 10.0. The CDI method is relatively simple and easy to perform. In addition, supports that have been activated by production imidazolylcarbamate groups are more stable to hydrolysis than those activated by the NHS method. A CDI-activated support is stable when stored in dry dioxane, with a half-life of greater than 14 weeks. Another advantage of this method is that the amide linkages formed by this technique (as well as those created by the NHS method) are more stable than the isourea linkages obtained by CNBr immobilization. One disadvantage of the CDI method is that it tends to produce ligands with a lower activity than alternative techniques (e.g., reductive amination) (Figure 14).

e. Other methods; One example is the use of cyanuric chloride (or 2,4,6-trichlorotriazine) to activate hydroxyl- or amine-containing supports for ligand attachment. Cyanuric chloride has been widely employed as a cross-linking agent and as a reagent for protein modification. It has three reactive acyl-like chlorines, each of which has a different chemical reactivity. The first chlorine is reactive toward hydroxyls and amines at 4°C and pH 9. After the first chlorine reacts, the second requires a slightly higher temperature for its reaction (20°C), and the third chlorine needs an even higher temperature (80°C). Other techniques for amine-containing ligands include the azalactone, divinylsulfone, bisoxylane, ethyldimethylaminopropyl carbodiimide, and tresyl chloride-tosyl chloride methods. Some of these methods are specific for certain supports (e.g., the azalactone method),

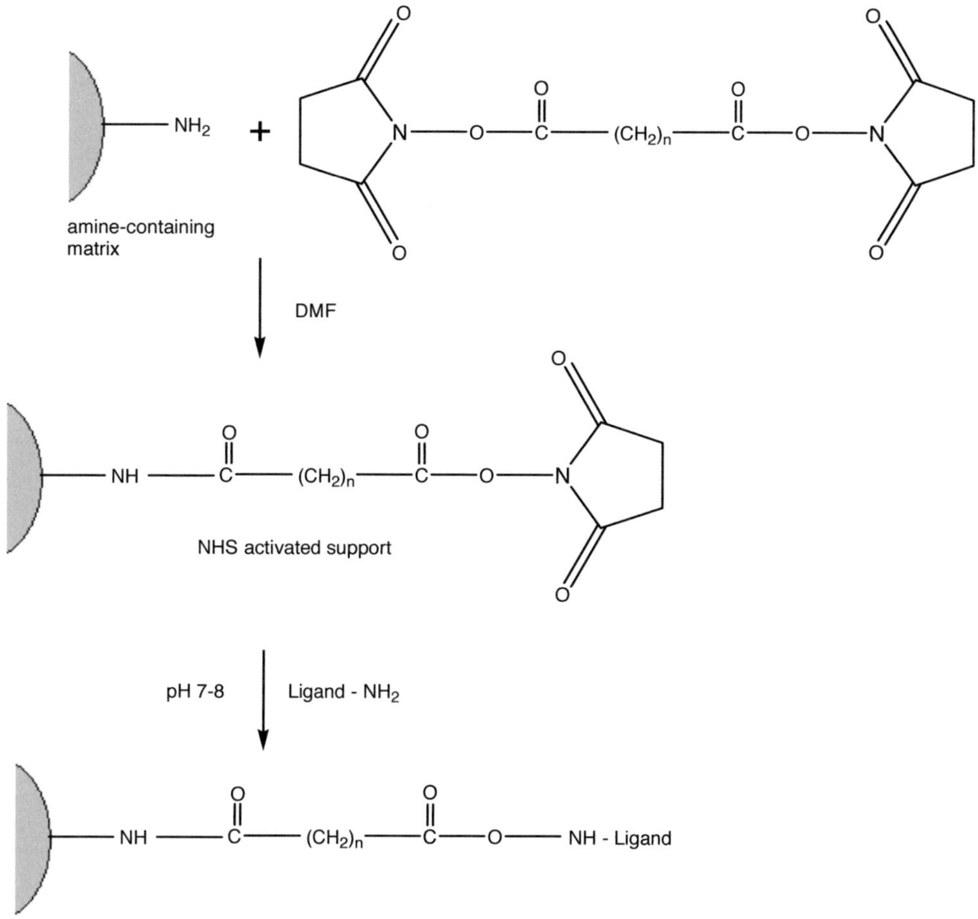

Figure 13. *N*-Hydroxysuccinimide immobilization method [8]

while others can be used with a variety of materials. The final selection among these approaches will often depend on the type of ligand being immobilized, the support desired for this ligand, and the conditions that can be tolerated by both the ligand and support during the immobilization process.

2. Sulfhydryl-Reactive Methods; The use of sulfhydryl groups on ligands is another approach for preparing affinity supports. If a ligand has a free sulfhydryl group on its surface, using this group is advantageous, since it often provides site-specific immobilization and a cleavable product. If the ligand is a protein or peptide that has no free sulfhydryl groups but that does have a disulfide bond, this bond can be reduced to allow ligand attachment. It is also possible to introduce sulfhydryl groups on a ligand by thiolating aminesor carboxyl groups. A support can be activated in several ways for the immobilization of ligands through sulfhydryl groups. Unlike amine-reactive methods,

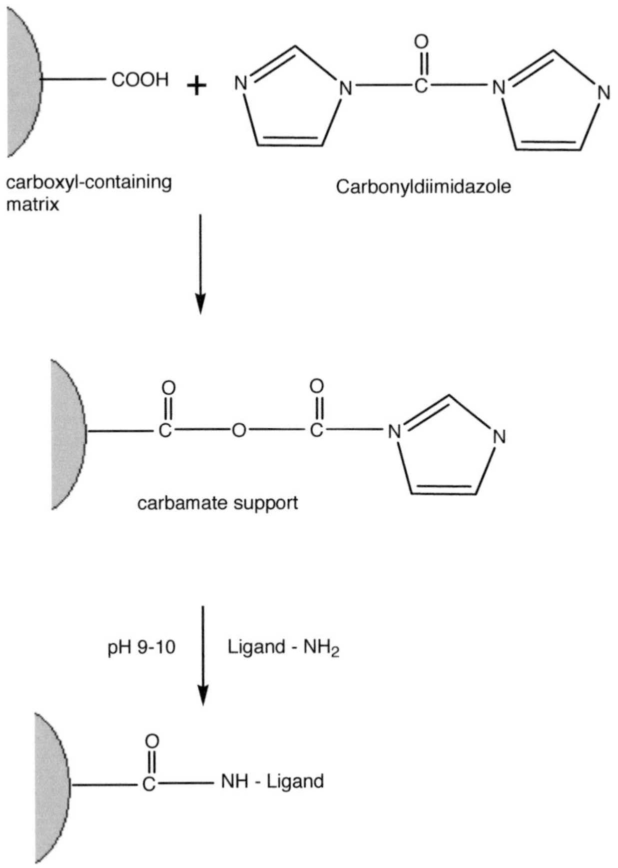

Figure 14. The carbonyldiimidazole immobilization method [8]

where hydroxyl groups on the support are generally used, most sulfhydryl-reactive methods require the introduction of an amine,carboxyl group, or some other intermediate site onto the support. For example, silica cannot be used directly with sulfhydryl-reactive methods but must be reacted with aminopropyltriethoxysilane or mercaptopropyltrimethoxysilane to convert it into a suitable form.There are various approaches that can be used to immobilize ligands with sulfhydryl groups. The following subsections examine some of these techniques, including the haloacetyl, maleimide, and pyridyl disulfide methods.

a. Haloacetyl Method; The haloacetyl method uses supports that contain iodoacetyl or bromoacetyl groups for the immobilization of ligands through sulfhydryl residues. These supports are usually prepared via the reaction of an amine-containing material with iodoacetic or bromoacetic acid in the presence of ethyldimethylaminopropyl carbodiimide (EDC) at pH 4 to 5. EDC reacts with the carboxylic acid in iodo- or bromoacetic acid to form a reactive ester, which can react with primary amine groups on the support. The second part of this process involves combination of the haloacetyl-activated support with a ligand containing a sulfhydryl group. This reaction proceeds by nucleophilic substitution and produces a thioether. The resulting bond is comparable to an amide linkage in stability. Although the reactivity of haloacetyl-activated supports toward sulfhydryls is relatively selective, these can react with methionine, histidine, or tyrosine under appropriate conditions. If the immobilization is carried out above pH 8, amines can also react with these supports.

b. Maleimide Method; Maleimides are another group of reagents employed for the selective coupling of a ligand through sulfhydryl groups. These tend to be more selective than a haloacetyl for such a reaction. The activation of a support with a maleimide is accomplished by using a homobifunctional or heterobifunctional cross-linking agent. One agent employed for this purpose is bis-maleimidohexane (BMH), which is a homobifunctional cross-linker with a maleimide group on both ends. The first of these groups can react with a support that has a sulfhydryl group. After the excess BMH has been washed away, the maleimide at the other end can react with a sulfhydryl group on a ligand.

c. Pyridyl Disulfide Method ; Pyridyl disulfide (or 2,2′-dipyridyldisulfide) is a homobifunctional cross-linking agent used for immobilizing ligands with sulfhydryl groups to supports that contain sulfhydryls on their surface. Activation of the support is accomplished by disulfide exchange between the sulfhydryl groups on the support and pyridyl disulfide, giving rise to the release of pyridyl-2-thione.

d. Other Methods; Divinylsulfone (DVS) can be used to activate a hydroxylcontaining support by introducing a reactive vinylsulfonyl group on its surface at pH 10 to11. This support can then be reacted with ligands that contain sulfhydryl, amine, or hydroxyl groups, with the rate of this reaction following the order –SH > –NH > –OH. Although the resulting bond for a sulfhydryl group is labile, the linkage for amine-containing ligands is more stable.

3. Hydroxyl-Reactive Methods; A number of methods have been used to couple ligands through hydroxyl groups. However, unlike many amine- and sulfhydryl-reactive methods, techniques for hydroxyl-containing ligands are not that selective. For example, the divinylsulfone method can be used for coupling an amine-, sulfhydryl-, or hydroxyl-containing ligand. Many supports used in affinity chromatography already contain hydroxyl groups on their surface. One way for the activation of these groups is to introduce bisoxirane (epoxy) groups. The most frequently used oxirane for this purpose is 1,4-butanediol diglycidyl ether, which contains two epoxy groups. One of the epoxy groups can react with the hydroxyl groups on a support while the other is used for coupling ligands containing sulfhydryl, amine, the reactivity of the terminal epoxide to other groups follows the order –SH > –NH > –OH. Strong alkaline conditions (pH 11) allow coupling by this method through hydroxyl groups, while amines and sulfhydryl groups can react at a lower pH (pH 7 to 8). Cyanuric chloride is another agent used for attaching a hydroxyl-containing ligand to a support. In this method, this can only be used effectively in the absence of amine groups due to the higher reactivity of these groups. As before-mentioned, divinylsulfone can be used for coupling hydroxyl-containing ligands. This, however, is not usually performed if the immobilized ligand is present at a pH higher than 9 to 10.

4. Carbonyl-Reactive Methods; Although most immobilization techniques involve coupling ligands through amine or sulfhydryl groups, the large number of such groups can create a problem with improper orientation or multipoint attachment. This can be avoided by using alternative groups that occur only in specific locations on the ligand. One example is the immobilization of antibodies through their carbohydrate residues. To use the carbohydrate groups of an antibody (or any other glycoprotein) for immobilization, these groups must first be oxidized to form reactive aldehyde groups. This can be accomplished by enzymatic treatment; however, it is usually performed through mild treatment with periodate. These aldehyde groups are then reacted with a support containing amine or hydrazide groups for ligand immobilization. This approach has been used not only for antibodies but also for glycoenzymes, RNA, and sugars. Supports with amine groups can be used for coupling aldehyde-containing ligands by reductive amination. Hydrazide-activated supports can also be employed for immobilizing ligands with aldehyde groups. Such supports can be prepared by formation aldehyde groups on the support and reaction of these with an excess of a dihydrazide (e.g., oxalic or adipic dihydrazide). An advantage of using a hydrazide-activated support is that no reducing agent is needed to stabilize the linkage between the ligand and support, as is required in reductive amination.

5. Carboxyl-Reactive Methods; There are currently no activated supports that react specifically with a ligand containing carboxyl groups. This is a result of the low nucleophilicity of carboxyl groups in an aqueous solution. However, there are reagents that will react with carboxylic acids and allow them to be activated for ligand attachment. 1-Ethyl-3-(dimethylaminopropyl) carbodiimide is an example of such a reagent. One problem of this process is that severe cross-linking is possible, since amine groups as well as carboxyl groups can react if excess EDC is present. In addition, the activated derivative formed, O-acylisourea, is not stable in an aqueous environment. This means that the activated ligand must be used immediately for immobilization without further purification.

6. Other Immobilization Techniques; Along with noncovalent and covalent immobilization methods, other techniques have been developed for the preperation of affinity supports. Such methods include entrapment, molecular imprinting, and the use of the ligands as both the support and stationary phase. Although these methods are not as common as the approaches already examined, they have important advantages in some applications [8].

Activation methods which are used in affinity chromatography can be summarized as follow:

Amine groups :

- Cyanogen bromide (CNBr) method
- Schiff base (reductive amination) method
- N-hydroxysuccinimide (NHS) method

http://dx.doi.org/10.5772/55781

- Carbonyldiimidazole (CDI) method
- Cyanuric chloride method
- Azalactone method (for Emphaze supports)
- Divinylsulfone (DVS)
- Epoxy (bisoxirane) method
- Ethyl dimethylaminopropyl carbodiimide (EDC)
- method
- Tresyl chloride/tosyl chloride method

Sulfhydryl groups:

- Azalactone method (for Emphaze supports)
- Divinylsulfone method
- Epoxy (bisoxirane) method
- Iodoacetyl/bromoacetyl method
- Maleimide method
- Pyridyl disulfide method
- TNB-thiol method
- Tresyl chloride/tosyl chloride method

Hydroxyl groups:

- Cyanuric chloride method
- Divinylsulfone method
- Epoxy (bisoxirane) method

Aldehyde groups

- Hydrazide method

Carboxyl groups

1.4. Elution

Elution is one of the critical step for successful separation. Sample application in affinity chromatography is performed usually by injection or application in the presence of mobile phase which is prepared in appropriate pH, ionic strength and solvent composition for solute-ligand binding. This solvent is referred as application buffer [8]. In the presence of application buffer, compounds which are complementary to the affinity ligand will bind while the other solutes in the sample will tend to pass through the column as nonretained compounds. After

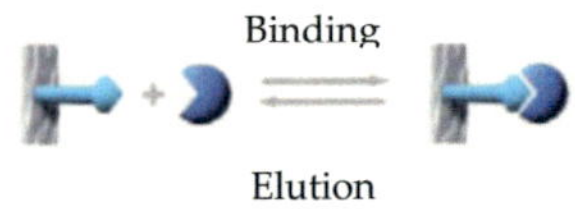

$$L + T \rightleftharpoons LT$$

$$K_D = \frac{[L][T]}{[LT]}$$

all nonretained components are washed off the column, the retained solute or together with ligand as solute-ligand complex can be eluted by applying a solvent. This solvent which is referred as elution buffer represents the strong mobile phase for the column. Later all the interest solutes are eluted from the column, regeneration is performed by elution with application buffer and the column is allowed to regenerate prior to the next sample application [4,8,21]. Step gradient elution or in other word on/off elution method is the most common method employed for affinity chromatography. Figure 15 shows the typical separation in affinity using step gradient elution.

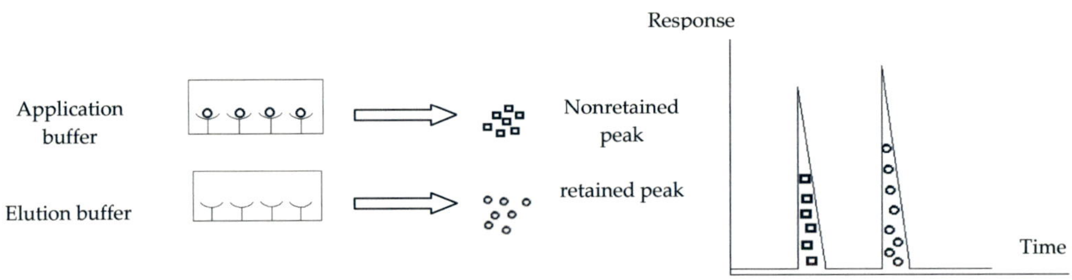

Figure 15. Typical separation in affinity using step gradient elution.

Step elution mode is employed if the ligans have high affinity for the target molecule. It is also possible to use isocratic elution in affinity chromatography. This elution mode generally selected if the target molecule and ligand have weak interaction. This approach is known as **Weak Affinity Chromatography** or **Dynamic Affinity Chromatography** [8,21].

In affinity chromatography there are many factors such as strength of solute-ligand interaction, the amount of immobilized ligand present and the kinetics of solute-ligand association and dissociation which have important influences on retention and elution of the compound. The reaction between the target protein (T) and ligand (L) on the other word binding (adsorption) and elution (desorption) process can be explain by following equation in case of a target protein has single site binding to a ligand [1,4,21].

K_D is the equilibrium dissociation constant

[L] is the concentration of free ligand

[T] is the concentration of free target

[LT] is the concentration of the ligand/target complex

The equation that is placed below explains the bound target-total target ratio. In order to achieve successful binding the ratio should be near 1 in this equation.

$$\frac{\text{Bound target}}{\text{Total target}} \approx \frac{L_0}{K_{D+L_0}}$$

L_0 is the concentration of ligand (usually 10^{-4} - 10^{-2}M)

K_D is the equilibrium dissociation constant

K_D can be changed by pH, ionic strength, temperature and other parameters. Therefore these parameters can be used to control the binding and elution efficiency of the reaction [1,4,22]

Obtaining stable biomolecules in high yield and purity is aimed for elution process. Elution is achieved by reducing the association constant of the ligand-solute interaction. Biospecific or non-specific elution can be utilized. Biospecific elution is based on solute displacement from the column by addition of molecule that acts as a competing agent. Two different types of biospecific elution can be applied for elution. In first method, normal role elution, molecule competes with the ligand for binding the desired solute. In second type of biospecific elution, reversed role elution, molecule competes with desired solute for binding the ligand [13]. The main advantage of biospecific elution is that a target can be gently removed from the column. However this elution is slow and generally results in broad solute peaks. Additionally competing agent needs to be removed from the eluted solvent therefore usage is limited. Another disadvantage especially in analytical application is need to use a competing agent that does not produce a large background signal under the conditions used for analyte detection [8]. Non-specific elution is performed by changing solvent conditions like pH, ionic strength and polarity. High concentration of chaotropic salts (NaCl, $MgCl_2$ or LiCl), denaturating agents and detergents (guanidine hydrochloride, sodium dodecyl sulfate and urea) can be used. Organic solvents can be used especially for the elution of low molecular weight compounds [13]. Alteration in structure of the solute or ligand which leads to a lower association constant and lower solute retention is provided by nonspecific elution [8]. Non-specific elution is faster than specific elution but there is a risk for denaturation of solute. The conditions which are applied for the elution may be too hard for column. If this is not considered it may result in long column regeneration times or irreversible loss of ligand activity [8].

For biospecific elution solvent is selected according to the type of target and ligand. The solvent usually has a pH and ionic composition similar to the application buffer but contains a competing agent. Reversed role elution is generally preferred when the target is a small compound while the normal elution is often used for isolation of macromolecules. Readily available in an inexpensive form and be soluble in the elution buffer are desired properties for competing agent in reverse role elution. In reversed-role elution it must be possible to remove the competing agent from the target when the affinity column is used for purification [8].

A wide range of mobile phase additives can be used in non-specific elution. In this elution nature of the target-ligand interaction is changed. This can be achieved by several ways such as altering pH of the targets and ligands that interact by weakly acidic or basic groups. Changing pH can lead to the alteration in the conformation of the target or ligand. Either increasing or decreasing of pH value can be used for this purpose but decreasing of pH is commonly preferred. Irreversibly denaturation of target, ligand or support may occur in this step. Collection of the eluted target in a neutral pH buffer and regeneration the column as soon as possible after the elution step can overcome this problem [8]. Changes in ionic strength induced by high salt solutions are a second way for nonspecific elution. Disruption of ionic bonds can be achieved by this method but hydrophobic interactions are promoted. Chaotropic salts (NaCl, $MgCl_2$ or LiCl) are useful for altering retention of targets. They distrupt the stability of water and interfere with hydrophobic interactions [8,23]. The main advantage of using either chaotropic salt of a change in ionic strength is that this usually leads to gentle elution of the target in an active form [8]. Denaturating agents such as urea, guanidine hydrochloride and sodium dodecyl sulfate which dissociate hydrogen bonds can also be used for elution. Sodium dodecyl sulphate (SDS) contains both hydrophobic and strong ionogenic groups and binding to hydrophobic regions results in a layering of negative charges on the protein's surface, causing irreversible unfolding of the structure. The denaturating effect of these solutions limits their usage. They should be only used in analytical applications if the ligand is quite stable or in preparative applications if both the ligand and target are relatively stable and it is enable to recover their activity after such elution [8,23]. Organic solvents in the mobile phase are also used in some cases such as using of 1-propanol in chiral affinity separations in order to improve solute retention and produce narrow peaks for good resolution. Polyols like ethylene glycol are also utilized in affinity separations [8]. In order to select the elution buffer several approaches can be followed. However the best way is that the buffer should be selected based on information in the literature, structure of the ligand, target and past experiences with these substances [8].

2. Types of affinity chromatography

The type of ligand can be used to divide affinity techniques into various subcategories such as lectin, immunoaffinity, dye ligand etc. These techniques are placed as below [1].

2.1. Immunoaffinity chromatography

Immunoaffinity chromatography is one of the most popular techniques of affinity derivatived method and it enables to produce ligands in case the ligand required is not available [7]. In this technique, stationary phase comprises of an antibody or antibody-related agent [1]. It is possible to isolate variable subtances using this technique due to high specifity of antibodies [1]. It is reported that immunoaffinity chromatography may be used for natural food contaminants such as aflatoxins, fumonisins and ochratoxins [11].

On the purpose of purification using antibodies as ligands, antibodies initially are immobilized on a support. In order to,bind the ligand on the surface of the support properly, protein A and G are usually used as a bridge which provides enough space for the ligand-protein binding.

Columns, dialysis membranes, capillaries or beads may be used in immunoaffinity application which is a non-covalent, irreversible purification process based on highly specific interactions between analyte and antibody [11].

Initially, the antibodies should be purified prior to prepare the immunoaffinity column. Precipitation with ammonium sulfate, ion-exchange chromatography, gel filtration chraomatography or affinity chromatography may be employed with the aim of antibody purification. Activated beads which are coated with bacterial proteins A or G may be used as the support material. Some parameters may be changed for the elution of the sample solution for example the ionic conditions of mobile phase may be changed or chaotropic buffers may be used [11].

Both small and large analytes can be determined using direct detection in IAC. Additionally it is possible to use this technique either separately or in combination with other chromatographic techniques [1]. If this technique is performed as part of HPLC system the method can be referred as high performance immunoaffinity chromatography.

Immunoaffinity chromatography is probably the most highly specific of all forms of bioaffinity chromatography. However this technique has some disadvantages such as: this technique relatively high cost, leakage of ligands may accur from the column and sometimes the desorption procedure results in partial denaturation of the bound protein [24].

2.2. Protein A or protein G affinity chromatography

Protein A is produced by *Staphylococcus aureus* while protein G is of group G *Streptococci*. These ligands are capable of binding to many types of immunoglobulins at around neutral pH and they dissociate in a buffer with a lower pH [1]. Protein A binds to the immunoglobulin G (IgG) obtained by human and other mammalian species with high specificity and affinity. In some cases protein G may be used instead of protein A [24]. These two ligands differ in their ability to bind to antibodies from different species and classes. Strong specifity and binding properties to immunoglobulins of protein A and protein G serve them as good ligands for the seperation of immunoglobulins. Protein A and protein G have use as secondary ligands for the adsorption of antibosies onto the support material in immunoaffinity applications. This method may also be employed in case high antibody activity or replacement of the antibodies in the affinity chromatography is needed [1].

2.3. Lectin affinity chromatography

Lectins which are non-immun proteins are produced by plants, vertebrates and invertebrates. Especially various plant seeds synthesize high levels of lectins [24]. Certain types of carbohydrate residues may be seperated via this method due all lectins have the ability to recognize and bind these types of compounds. Mostly used lectins for affinity columns are concanavalin A, soybean lectin and wheat germ agglutinin [1, 24]. Concanavalin A is specific for α-D-mannose and α-D-glucose residues while wheat germ agglutinin binds to D-*N*-acetyl-glucosamine. Lectins which are commonly used for the isolation of compunds containing carbohydrates such as polysaccharides, glycoproteins and glycolipids in affinity chromatography are given in Table 4. [1].

Lectin	Source	Sugar specificity	Eluting sugar
Con A	Jack bean seeds	α- $_D$-mannose, α- $_D$ -glucose	α - $_D$ -methyl mannose
WG A	Wheat germ	*N*-acetyl-β- $_D$ -glucosamine	*N*-acetyl-β- $_D$ -glucosamine
PSA	Peas	α- $_D$ -mannose	α- $_D$ -methyl mannose
LEL	Tomato	*N*-acetyl-β- $_D$ -glucosamine	*N*-acetyl-β- $_D$ -glucosamine
STL	Potato tubers	*N*-acetyl-β- $_D$ -glucosamine	*N*-acetyl-β- $_D$ -glucosamine
PHA	Red kidney bean	*N*-acetyl-β- $_D$ -glucosamine	*N*-acetyl-β- $_D$ -glucosamine
ELB	Elderberry bark	Sialic acid or *N*-acetyl-β- $_D$ -glucosamine	Lactose
GNL	Snowdrop bulbs	α -1→ 3 mannose	α -methyl mannose
AAA	Freshwater eel	α – $_L$-fucose	$_L$ -fucose

Table 4. Some lectins which are commonly used in affinity chromatography

Enzymes, inhibitors, cofactors, nucleic acids, hormones or cell chromatography can also be utilized as ligands in bioaffinity chromatography types. Examples of these methods include Receptor Affinity Chromatography and DNA Affinity Chromatography [21].

2.4. Dye-ligand affinity chromatography

Development of the dye-ligand affinity chromatography could be attributed to observation of some proteins irregular elution characteristics during fractionation on gel filtration column in presence of blue dextran. Blue dextran consists of a triazine dye (cibacron blue F3G-A) covalently linked to high molecular mass dextran. Some proteins bind triazine dye and this allows to its use as an affinity adsorbent by immobilization [24]. This method is especially popular tool for enzyme and protein purification [21]. Dye-ligand adsorbents are of interest due to inexpensiveness, ease of availablity and immobilization process. These adsorbents may be used in analytical, preparative analysis and large scale studies. Although dye-ligand affinity technique for pharmaceuticals may be preferred owing to these advantages, concerns about leakage and toxicity has stopped its use. Therefore proteins purified using this technique is convenient for analytical or technical uses. Procion Red HE3b, Red A, Cibacron Blue F3G-A are some examples of dye-ligands which are used for purification [9].

2.5. Metal-chelate affinity chromatography (Immobilized-metal (Ion) affinity chromatography)

In 1970s, first application of metal-chelate affinity chromatography which is later named as "immobilized-metal (ion) affinity chromatography (IMAC) was perfomed. Metal-chelate chromatography technique exploits selective interactions and affinity between transition metal immobilized on a solid support (resin) via a metal chelator and amino acid residues which act as electron donors in the protein of interest [25-26]. As well as aromatic and heterocyclic compounds, proteins such as histidine, tyrosine, tyriptophane and phenylalanine posses affinity to transition metals which form complexes with compounds rich in electrons [25,27].

Among these amino acids histidine is the most commonly used one. Attachment of histidine tags to the recombinant proteins polypeptides is the most known development in the field of IMAC. Histidine and other metal affinity tags are widely used for protein purification [26]. Adsorbents may be prepared by binding chelators onto the surface and metals to the chelators. Free coordination sites of the metal ions are needed for the analyte to bind to metal ions [25].

Zn^{2+}, Ni^{2+} and Cu^{2+} are the most commonly used metal ions. Basic groups on protein surfaces especially the side chain of hisitidine residues, are attracted to the metal ions to form a weak coordinate bonds [24].

The metal ions in the class of hard Lewis acids such as K^+, Ca^{2+}, Mg^{2+}, Fe^{3+}; soft Lewis acids such as Ag^+, Cu^+ and transition metals classified as borderline acids (Co^{2+}, Cu^{2+}, Ni^{2+}) may be used in IMAC applications, especially Ni^{2+} which has six coordination sites and electrochemical stability. The affinity of the metals may be predicted according to the principles of soft acids and bases which is the theory explaining one of the two atoms attached acts as a Lewis acid and the other as a Lewis base. Ligands with oxygen (e.g. carboxylate), aliphatic nitrogen (e.g. asparagin, glutamine) and phosphor (phosphorylated amino acids) are hard Lewis bases, as the ones with sulfur (e.g. cysteine) are soft bases and those with aromatic nitrogen (e.g. histidine, tryptophane) are borderline bases. In case Cu(II), Ni(II), Co(II) or Zn(II) ions are the ions used in the IMAC, the target amino acids on the protein surface are imidazolyl, thiol and indolyl groups; as carboxyl and phosphate groups are of that in case of the use of Fe(III) and Mg(II). Histidine, tryptophane and cysteine are accepted to be the most important amino acids for IMAC due to their strong affinity to metal ions and the retention times. It is reported that histidine residues attached to the protein surface significantly change the retention time of the protein of interest [26].

Chelating Compound	Coordination	Metal Ions
Aminohydroxamic acid	bidentate	Fe(III)
Salicylaldehyde	bidentate	Cu(III)
8-Hydroxy-quinoline (8-HQ)	bidentate	Al(III), Fe(III), Yb(III)
Iminodiacetic acid (IDA)	tridentate	Cu(II), Zn(II), Ni(II), Co(II)
Dipicolylamine (DPA)	tridentate	Zn(II), Ni(II)
ortho-phosphoserine (OPS)	tridentate	Fe(III), Al(III), Ca(II), Yb(III)
N-(2-pyridylmethyl)aminoacetate	tridentate	Cu(II)
2,6-Diaminomethylpyridine	tridentate	Cu(II)
Nitrilotriacetic acid (NTA)	tetradentate	Ni(II)
Carboxymethylated aspartic acid (CM-Asp)	tetradentate	Ca(II), Co(II)
N,N,N'-tri(carboxymethyl)ethylenediamine (TED)	pentadentate	Cu(II), Zn(II)

Table 5. Some examples of chelating compouns used in IMAC [28]

Multidentate chelating compounds are widely used in order to strengthen the complex which is comprised of chelator, metal ion and protein. Different length of spacers is used to bind the chelator onto the surface of the support. Type of the chelator influences the strength of the chelation and retention power, for instance metal binds to the nitrogen atom and two carboxylate oxygens and reveals three free sites in case of tridentate iminodiacetic acid (IDA); tetradentate nitrilotriacetic acid (NTA) binds the metal by an additional carboxylate oxygen and this provides stronger chelation, but a weaker retention power. IDA is the chelator which is commonly used in the applications of IMAC. Although most of the chelators are carboxymethylated amines, there also some other compounds which are commonly used such as dye-resistant yellow 2KT, OPS and 8-HQ [26]. Some examples of chelating compounds are given in Table 5.

2.6. Boronate affinity chromatography

In case of use boronic acid or boronates as ligand of the affinity chromatography, this type of methods are called boronate affinity chromatography. Most of the boronate derivatives are known to bind compounds with cis-diol groups covalently at a pH above 8. Separation of glycoproteins from non glucoprotein structures is possible by boronate affinity method due to cis-diol groups of the sugars. For instance, this method may successfully performed to seperate glucohemoglabin and normal hemoglobin or to determine different types of glycoproteins in a sample [1].

There are many other chromatographic methods which are closely related to traditional affinity chromatography. For example **Analytical Affinity Chromatography** (**Quantitative Affinity Chromatography** or **Biointeraction Chromatography**) which is used as a tool for determination of solute-ligand interactions [21]. It is possible to investigate several biological systems, such as lectin/sugar, enzyme/inhibitor, protein/protein, DNA/protein interactions as well as binding of drugs or hormones to serum proteins with this technique. Thus competition of drugs with other drugs or endogenous compounds for protein binding sites may successfully be detected by this method. Either immobilized drugs or immobilized proteins may be used in the studies about drug-protein and hormone-protein binding, although protein-based columns which may be use for multiple experiments are more common [1,29]. The competition between two solutes for binding sites can also be examined by this method and this technique is known as **Frontal Affinity Chromatography** [21]. **Hydrophobic Interaction Chromatography** and **Thiophilic Adsorption** methods also related to affinity chromatography. Immobilized thiol groups are used as ligands in **Thiophilic Adsorption** (**Covalent/Chemisorption Chromatography**) in order to separation of sulfhydryl-containing peptides or proteins and mercurated polynucleotides. In **Hydrophobic Interaction Chromatography** short non-polar chain, such as those that were originally used as spacer arms on affinity supports provide binding with proteins, peptides and nucleic acids. **Chiral Liquid Chromatography** methods can be also considered as affinity based techniques [21]. These techniques are widely utilized in pharmaceutical industry and clinical chemistry for the separation of individual chiral forms of the drugs and the quantification of different chiral forms of drugs or their metabolites. Since most of the ligands used in affinity chromatography are chiral, they may be preffered as stationary phases for chiral seperations. Protein-based and carbohydrate-based ligands may be used as the stationary phases in the analysis of chiral compounds via HPLC [1]. Orosomucoid (α_1-acid glycoprotein), bovine serum albumine and ovomucoid (a glucoprotein of egg whites) are some examples of protein-based stationary phases, while cyclodextrins (especially β-cyclodextrin) are of carbohydrate-based stationary phases [1,29].

3. Affinity chromatography and drug discovery

There are a number of areas related to affinity chromatography that have also been of great interest in pharmaceutical and biomedical analysis. One such area is the use of affinity chromatography in drug discovery [12]. In drug discovery explaining the mechanism of action of bioactive compounds, which are used as pharmaceutical drugs and biologically active natural products, in the cells and the living body is important. For this pupose isolation and identification of target protein(s) for the bioactive compound are essential in understanding its function fully. Affinity chromatography is a useful method capable of isolating and identifying target molecules for a specific ligand, utilizing affinity between biomolecules such as antigen–antibody reactions, DNA hybridization, and enzyme–substrate interactions. Since the development of affinity chromatography in the early 1950s, various types of target proteins for bioactive compounds have been isolated and identified. Selected samples of the target proteins isolated by affinity chromatography are listed in Table 6. Since then, affinity chromatography has been gaining renewed attention as a widely applicable technique for the discovery the target proteins for bioactive compounds [30].

Bioactive compound	Molecular structure	Target protein
E-64 (cystein protease inhibitor)	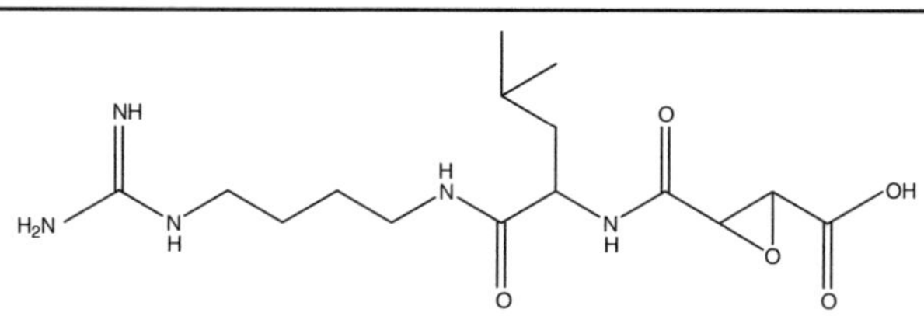	Cathepsin B, H, L
Indomethacin (antitumor drug)	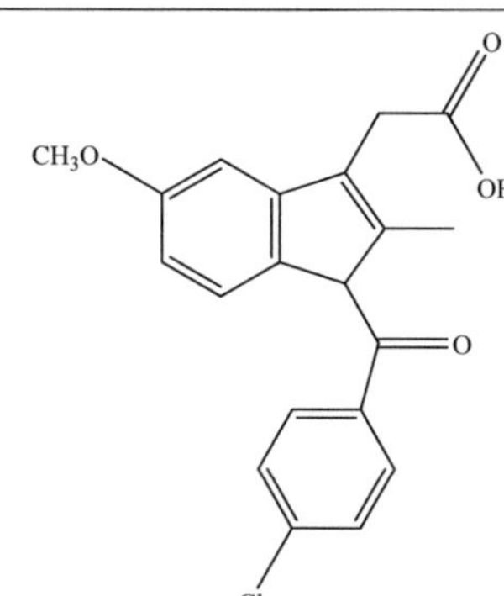	Glyoxalase 1

Bioactive compound	Molecular structure	Target protein
Methotrexate (anticancer drug)	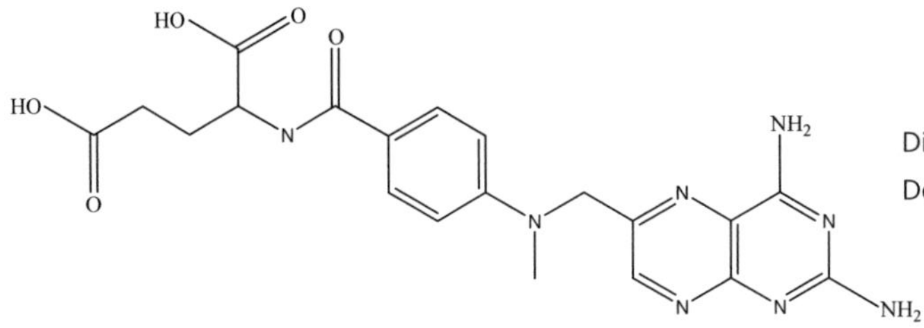	Dihydrofolate reductase Deoxycytidine kinase
Resvertatrol (health benefits including life span)	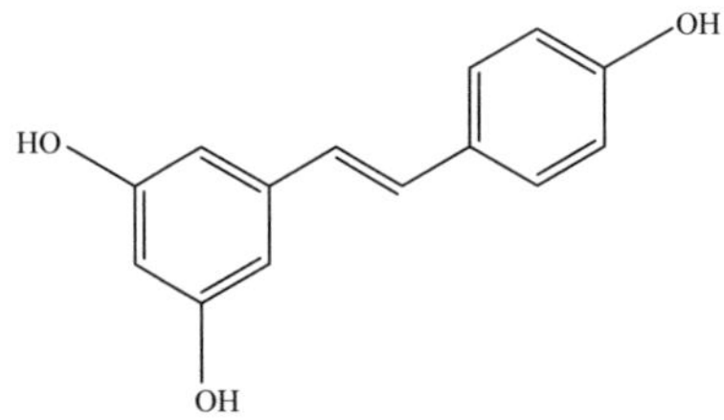	Tif1 (yeast elF4A)
Spliceostatin A (splicing inhibitor)	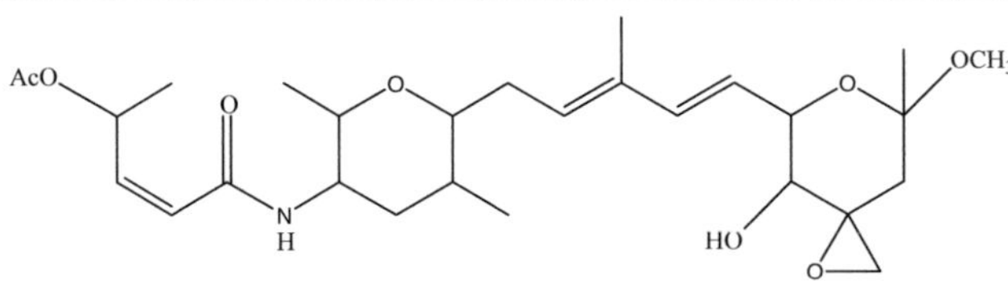	Splicing factor 3b
Thalidomide (sedative, anticancer drug)	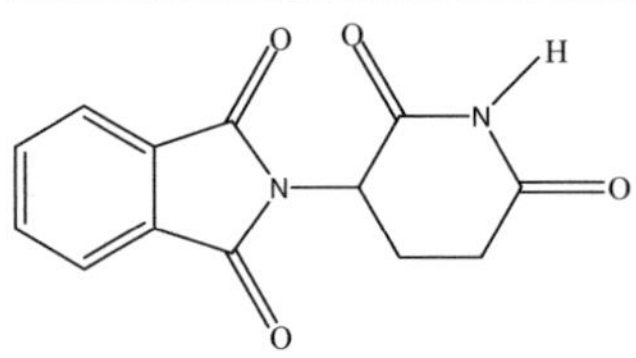	Cereblon
Wortmannin (PI 3-kinase inhibitor)	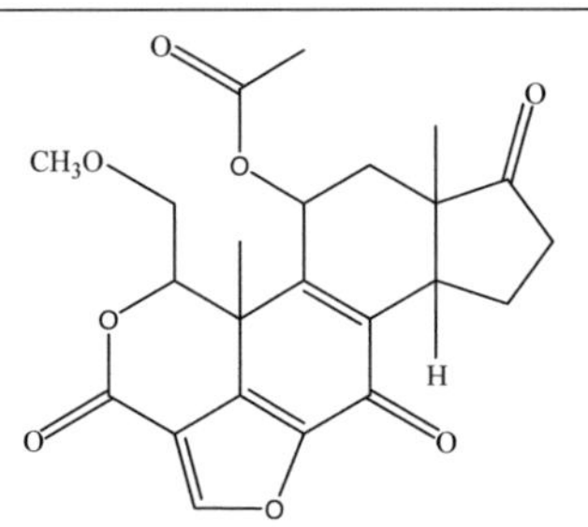	Polo-like kinase 1 Polo-like kinase 3

Table 6. Selected examples of the target protein(s) isolated and identified through affinity chromatography [30]

4. Conclusion

Liquid chromatographic techniques have been widely used in pharmaceutical and clinical laboratories. Reversed phase, ion exchange, size exclusion and normal phase chromatography are commonly used types of these techniques. Affinity chromatography which provides analysis of bioactive molecules based on their biological functions or individual structures has become increasingly important as another liquid chromatography technique [1,12]. Affinity chromatography is based on the simple principle that every biomolecule recognize another natural or artificial molecule such as enzyme and substrate or antibody and antigen [2,7]. In fact this technique is one of the oldest forms of liquid chromatography method [8]. The first use may be considered as the isolation of α-amylase by using an insoluble substrate, starch, in 1910 just three years after the discovery of chromatography by Tsewett [6,9]. The modern applications have started since 1960s with the creation of beaded agarose supports and the use of cyanogens bromide immobilization method. Since then, affinity chromatography has been gaining attention as a widely applicable technique for discovering the target proteins for bioactive compounds [3]. Up to the present time many different types of target proteins for bioactive compounds have been isolated and identified by affinity chromatography [21]. Today affinity chromatography is utilized as a valuable technique for the separation, purification and analysis of compounds present in complex samples and used in biochemistry, pharmaceutical science, clinical chemistry and environmental sciences. Application of affinity chromatography has significant advantages. The important one is that affinity chromatography involves many types of interactions between ligand and target such as steric effects, hydrogen bonding, ionic interactions, van der Waals forces, dipol-dipol interactions and even covalent bonds while other chromatographic techniques involve just one or a few of them. The combination of these multiple interactions leads to separation with high selectivity and retention in affinity chromatography [8]. However there are also several drawbacks in affinity chromatography system, such as non-specific binding of irrelevant proteins during affinity purification and chemical modification of bioactive compounds of interest used as ligands. These drawbacks limit its extensive application. After the completion of the Human Genome Project drug discovery research has focused on an approach that includes identification and characterization of molecular and cellular functions of a wide variety of proteins encoded by genomes. Thus, bioactive compounds become more important not only as therapeutic agents to treat diseases and disorders but also as useful chemical tools to examine their complex biological processes in vitro and in vivo [30]. Affinity chromatography which allows explaining the mechanism of action of bioactive compounds that are used as pharmaceutical drugs and biologically active natural products, therefore, has also significant importance in modern drug discovery [30].

Author details

Özlem Bahadir Acikara, Gülçin Saltan Çitoğlu, Serkan Özbilgin and Burçin Ergene

Ankara University, Faculty of Pharmacy, Department of Pharmacognosy, Ankara, Turkey

References

[1] Hage DS. Affinity chromatography: A review of clinical applications. Clinical Chemistry 1999;45(5) 593-615.

[2] Lowe CR. Combinatorial approaches to affinity chromatography. Current Opinion in Chemical Biology 2001;5 248-256.

[3] Urh M, Simpson D, Zhao K. Affinity chromatography: general methods. Methods in Enzymology 2009; 463:417-438.

[4] GE Healthcare. Affinity Chromatography Principles and Methods.http://www.gelifesciences.com/gehcls_ images/GELS/Related%20Content/Files/1334615650802/litdoc18102229_20120420130219.pdf (acessed 20 July 2012).

[5] Wilchek M, Chaiken I. An Overview of Affinity Chromatography. In: Bailon P, Ehrlich GK, Fung WJ, Berthold W. (ed.) Affinity chromatography Methods and Protocols. Totowa, New Jersey: Humana Press; 2000.

[6] Turkova J. Affinity chromatography. Journal of Chromatography Library Vol. 12. Amsterdam: Elsevier; 1978.

[7] Wilchek M, Miron T. Thirty years of affinity chromatography. Reactive&Functional Polymers 1999;41 263-268.

[8] Hage DS (ed), Cazes J (ed). Handbook of Affinity Chromatography Second Edition. Boca Raton, Florida: CRC Press; 2006.

[9] Labrou NE. Design and selection of ligands for affinity chromatography. Journal of Chromatography B 2003;790 67-78.

[10] Roque ACA, Lowe CR. Affinity Chromatography History, Perspectives, Limitations and Prospects. In: Zachorou M. (ed.) Affinity Chromatography: Methods and Protocols (Methods in Molecular Biology). Totowa, New Jersey: Humana Pres; 2010.

[11] Van Emon JM, Gerlach CL, Bowman K. Bioseperation and bioanalytical techniques in environmental monitoring. Journal of Chromatography B 1998;715 211-228.

[12] Hage DS, Anguizola JA, Bi C, Li R, Matsuda R, Papastavros E, Pfaunmiller E, Vargas J, Zheng X. Pharmaceutical and biomedical applications of affinity chromatography: Recent trends and developments. Journal of Pharmaceutical and Biomedical Analysis, (in press).

[13] Ayyar BV, Arora S, Murphy C, O'Kennedy R. Affinity chromatography as a tool for antibody purification. Methods 56:116-129, 2012.

[14] Kline T (ed). Handbook of Affinity Chromatography. New York: Marcel Dekker Inc; 1993.

[15] Matejtschuk P. Affinity Separations: A Practical Approach. New York: Oxford University Press; 1997.

[16] Oza MD, Meena R, Prasad K, Paul P, Siddhanta AK. Functional modification of agarose: A facile synthesis of a fluorescent agarose–guanine derivative. Carbohdrate polymers 2010; 81 878-884.

[17] Sphere Scientific Corporation Expert of advanced microspheres. http://www.spherescientific.com/PS-PSDVB.html (accessed 1 september 2011).

[18] Thu K, Chakraborty A, Saha BB, Ng KC. Thermo-physical properties of silica gel for adsorption desalination cycle. Applied Thermal Engineering (in pres).

[19] Zou H, Luo Q, Zhou D. Affinity membrane chromatography for the analysis and purification of proteins. Journal of Biochemical and Biophysical Methods 2001;49 199-240.

[20] Clonis YD. Affinity chromatography matures as bioinformatic and combinatorial tools develop. Journal of Chromatography A 2006;1101 1-24.

[21] Hage DS. Affinity Chromatography. In: Cazes J. (ed.) Encyclopedia of Chromatography, Third edition. Boca Raton, Florida: CRC Press; 2010.

[22] Walsh G (ed). Proteins Biochemistry and Biotechnology. Chichester, England: John Wiley Sons Ltd; 2002.

[23] Firer MA. Efficient elution of functional proteins in affinity chromatography. Journal of Biochemical and Biophysical Methods 2001; 49 433-442.

[24] Magdeldin S, Moser A. Affinity Chromatography: Principles and Applications. In: Magdeldin S. (ed.) Affinity Chromatography, Rijeka:InTech; 2012. http://www.intechopen.com/books/affinity-chromatography (accessed 1 September 2012).

[25] Fanou-Ayi L, Vijayalakshmi M. Metal-chelate affinity chromatography as a seperation tool. Biochemical Engineering III 1983;413 300-306.

[26] Ueda EKM, Gout PW, Morganti L. Current and prospective applications of metal ion-protein binding. Journal of Chromatography A 2003;988 1-23.

[27] Ji Z, Pinon DI, Miller LJ. Development of magnetic beads for rapid and efficient metal-chelate affinity purifications. Analytical Biochemistry 1996;240 197-201.

[28] Gaberc-Porecar V, Menart V. Perspectives of immobilized-metal affinity chromatography. Journal of Biochemical and Biophysical Methods 2001;49 335-360.

[29] Clarke W, Hage DS. Clinical applications of affinity chromatography. Separation&Purification Reviews 2003;32(1) 19-60.

[30] Sakamoto S, Hatakeyama M, Ito T, Handa H. Tools and methodologies capable of isolating and identifying a target molecule for a bioactive compound. Bioorganic and Medicinal Chemistry 2012;20 1990-2001.

Chapter 4

Chromatography in Bioactivity Analysis of Compounds

Sylwester Czaplicki

Additional information is available at the end of the chapter

http://dx.doi.org/10.5772/55620

1. Introduction

Chromatographic techniques have led to considerable development in mixture ingredient analysis. With the development of stationary-phase and high-pressure module technologies, new possibilities have emerged for the separation of complex systems whose components are characterised by similar structures and properties. Modern detection systems now allow detection and identification of individual ingredients. Data on their electrical properties, characteristic ways of molecule ionisation, and their ability to absorb or emit electromagnetic waves are used for this purpose (among others). The possibilities offered by chromatographic techniques are also used to isolate mixture ingredients for further analysis or to compose mixtures with predesigned properties. Appropriate tests preceded by isolation of analysed ingredients from the biological matrix are applied to determine the characteristics of components of analysed biological samples. In some circumstances this can be very arduous because the separation process itself is time- and cost-consuming or it is necessary to have appropriate infrastructure. Preparative-scale chromatographic separation can be conducted using the techniques of preparative thin-layer chromatography (prep-TLC), preparative column chromatography (prep-LC) as well as preparative-scale high-performance liquid chromatography (prep-HPLC). Isolated fractions or purified compounds can serve as material for determination of their biological properties. Currently, researchers are greatly interested in methods of analysis and identification for compounds which are antioxidants or inhibitors of specific transformations in the biochemistry of living organisms. Ingredients isolated from the material are purified by solid-phase extraction (SPE), thin-layer chromatography (TLC), preparative column chromatography (prep-LC) and preparative-scale high performance liquid chromatography (prep-HPLC). The obtained compounds or fractions are the subject of research to determine their biological properties in tests with DPPH, ABTS, AAPH radicals, the Folin-Ciocalteu reagent, etc. This procedure is cost-, time- and labour-consuming and the obtained results may be uncertain. This uncertainty is connected with the fact that during isolation, compounds are exposed to environmental factors such as oxygen, light, increased temperature. As a result the isolated compound may not have such a chemical structure as before isolation. This can cause a change in the biological activity of the examined substances. A combination of chromatographic ingredient separation methods with the detection of biochemical properties provides great possibilities for examination of the compounds present in complex biological systems. Methods have emerged which use the advantages of a solution to determine the ingredients of analysed mixtures with regard to both quality and the antioxidant activity detection. These methods have become very useful, e.g. during identification of the biological activity of plant extracts. The search for chemical compounds with desired biological properties by coupling chromatographic methods with biochemical detection has immense possibilities. The merits of this solution are currently used to an increasing extent.

This paper presents analysis methods for compounds to determine their general biological activity. Characteristics of the methods in model systems using different antioxidant reaction mechanisms are also described. Both colorimetric methods and those with fluorometric detection as well as chemiluminescence testing are included.

The main element in the paper is presentation of the possibilities for using the liquid chromatography technique for screening compounds with regard to their biological properties. Examples of different uses of chromatographic methods in on-line analysis of the bioactivity of mixture ingredients are also described.

2. Separation of components by means of gas chromatography

Chromatographic techniques are based on separation of substances between a stationary and a mobile phase. The mobile phase moves relative to the stationary one. Components of a mixture to be separated move together with the mobile phase due to their different interactions with the phases.

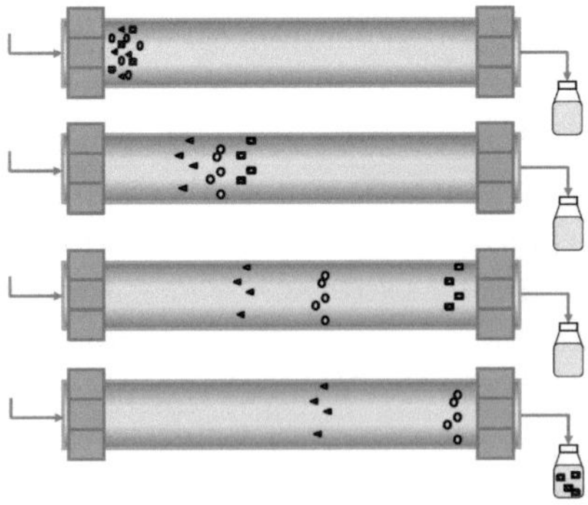

Figure 1. Separation of a compound on a chromatographic column

http://dx.doi.org/10.5772/55620

Depending on the technique applied, the stationary phase can be: column packing (column chromatography), thin layer of active substances put onto a plate (alumina or predominatingly silica gel) (thin-layer chromatography) or absorbent paper (paper chromatography). A mobile phase is a liquid which moves relative to the stationary phase. To be separable by this technique, components of a mixture must be soluble in the mobile phase. Depending on the interactions between the components and individual phases, the components move faster or more slowly together with the mobile phase along the stationary phase. The speed of movement depends on the strength of interactions between the components and each of the phases.

Such interactions cause sorption and desorption of the components from the mobile to the stationary phase. For the separation to be effective, the speeds of movement of components relative to the stationary phase must be different. The simplest solutions are based on free flow of solvent, but the system effectiveness can be improved by using pressurised flow. Upon leaving the chromatographic system, the separated components flow to detectors, where their specific properties are used to observe their presence, amount, and sometimes event to identify them. UV-Vis detectors, measuring absorbance of the solution which leaves a chromatographic column, are among the most widely-used devices. Components of a solution are observed as increased light absorption by the solution flowing through a quartz flow cell. With a detector of this type, it is possible to observe elution of individual mixture components on the "present-absent" basis. An analyst is presented with better capabilities when using a version of the detector with a photodiode matrix. Owing to a photodiode detector (PDA), it is possible to observe light absorption within the UV-Vis range at different wavelengths simultaneously. Moreover, it is possible to conduct observations of the absorption spectrum within a selected part of the UV and visible range. Owing to that property, it is possible to conduct simultaneous observation of elution of compounds which absorb light to a different extent at different wavelengths and, additionally, the detector is more selective. It is very important that identification of the separated compounds can be conducted based on the spectra. This is possible because individual compounds have specific absorption spectra depending on the bond structures and function groups. A less popular detection technique makes use of the ability of compounds to emit light. Fluorescence detectors are used where separated compounds specifically emit energy after their excitation. These detectors are highly selective and sensitive, which is essential when other mixture components are co-eluted with the substances being determined. In such cases, the detection parameters are set to make the excitation or emission wavelength match the analysed compounds. Mentioned detectors make use of the ability to absorb or emit light, but other detection techniques are also used depending on the properties of the analysed compounds. These include the following types of detectors: refractometric – the signal is measured as a change of the light refraction coefficient caused by optically active substances; electrochemical – recording a change of the electric potential; detector of dispersed light – measuring the intensity of dispersion of a laser beam by molecules of the substance being separated; mass detectors – analysing compounds following their ionisation [1]. Apart from those mentioned here, which are the most popular, other techniques of detection are also applied, with different selectivity and using different properties of the analyte and with different degrees of sample degradation. If liquid chromatography is used to obtain mixture components with a view to further analysis, a method of detection must be used which does not change the structure or properties of the compounds under analysis. The most popular one in such cases is a UV-Vis detector. Spots of substances being separated by paper or thin-layer chromatography are observed under visual or UV light, in their natural form or after transformation into a coloured compound. Both TLC and column chromatography are used in analysis of antioxidant compounds. In TLC, substances previously separated on the plate affect the intensity of colour of the radical placed on it. In liquid chromatography, tests of antioxidant activity of different components can be performed after they are separated in a pure state, by performing post-column off-line reactions or during the chromatographic separation on-line.

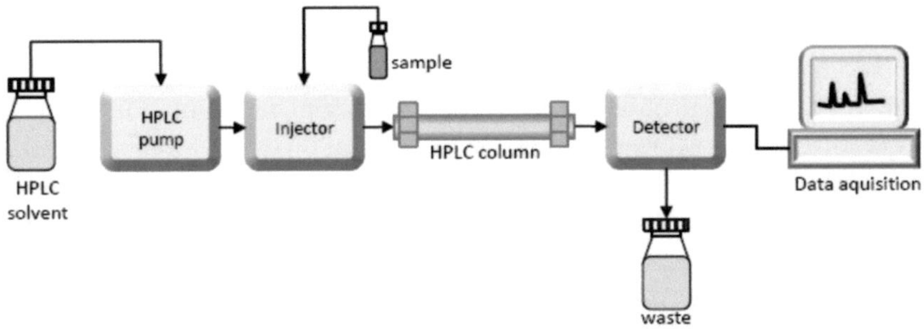

Figure 2. Schematic diagram of the High Performance Liquid Chromatography (HPLC) system

3. In vitro analysis of biological activity of substances.

In search of bioactive substances, researchers have directed their interest towards substances found in plants. Parts of plants which have been used in natural medicine have proved to be a rich source of bioactive compounds; however, to make use of them, they have to be isolated and their properties determined. Using selective techniques of extraction has resulted in obtaining concentrated preparations of bioactive substances. To achieve comprehensive knowledge of their properties, it was necessary to develop methods of isolation of individual components and testing these methods. This could be done with chromatographic techniques. Isolated compounds were tested in order to show which of them (and to what extent) are responsible for bioactivity of plant preparations from which they were obtained. Due to the fact that many of the substances have the opposite effect, it is frequently impossible to use extracts without isolating individual compounds.

In vitro tests, used in evaluation of antioxidant properties make use of the ability of antioxidants to quench free radicals. Based on this mechanism, the methods are divided into two groups: SET – single electron transfer, and HAT – hydrogen atom transfer. Reactions with antioxidants in assays with the DPPH radical, ABTS and the Folin-Ciocalteu reagent both operate according to the SET and HAT mechanism. Due to the kinetics of the reaction, they are included in the group of SET assays. The HAT mechanism is of lesser importance in those assays [2]. SET assays include: DPPH, TEAC, FRAP, CUPRAC, DMPD, Folin-Ciocalteu; HAT assays include: ORAC, TRAP, CBA, β-carotene – linoleic model system. Those classified as "other" in literature include: cellular antioxidant activity (CAA), chemiluminescence, electrochemiluminescence, Total Oxyradical Scavenging Capacity Assay (TOSCA) and others [3].

3.1. Single Electron Transfer (SET) methods

3.1.1. 22-diphenyl-1-picrylhydrazyl (DPPH•) assay

Analysis of antioxidant properties relative to the DPPH• radical involves observation of colour disappearance in the radical solution in the presence of the solution under analysis which contains antioxidants. A solution of extract under analysis is introduced to the environment containing the DPPH• radical at a specific concentration. A methanol solution of the DPPH• radical is purple, while a reaction with antioxidants turns its colour into yellow. Colorimetric comparison of the absorbance of the radical solution and a solution containing an analysed sample enables one to make calculations and to express activity as the percent of inhibition (IP) or the number of moles of a radical that can be neutralised by a specific amount of the analysed substance (mmol/g). In another approach, a range of assays are conducted with different concentrations of the analysed substance to determine its amount which inactivates half of the radical in the test solution (EC_{50}). The duration of such a test depends on the reaction rate and observations are carried out until the absorbance of the test solution does not change [4]. If the solution contains substances whose absorbance disturbs the measurement, the concentration of DPPH• radical is measured directly with the use of electron paramagnetic resonance (EPR) spectroscopy.

The disadvantage of the method is the fact that there are numerous modifications in the literature. These include using radical solutions at different concentrations, different reaction times and sometimes even a different reaction environment. The effect of exposure to light, the presence of oxygen, pH and the type of solution on the stability of the DPPH• radical have been studied by Ozcelik and co-workers [5]. The study methods developed so far employ radical concentrations ranging from $6*10^{-5}$ to $2.0*10^{-4}$ mol/dm^3 [6, 7, 8, 9, 10]. Depending on the method, absorbance of the reaction mixture is measured at the wavelength ranging from 515 nm to 550 [11, 12] after a specified reaction time, e.g. 5 min [13], 10 min [14], 16 min. [9], 20 min [7] 30 min [15], 60 min [16]. As Sánchez-Moreno and coworkers [17] found in their study, the time after which the absorbance of the analysed solution takes place depends on various factors, e.g. on the antioxidant concentration. For individual standards, they determined the time needed to reach the plateau at an antioxidant concentration of EC_{50}, i.e. such that is necessary to achieve 50% inactivation of the DPPH• radical. In order to make the results comparable, researchers frequently express them as Trolox equivalent [18]. The application of the assay is limited by the fact that the DPPH• radical is non-polar and as such, it is soluble in organic solvents. Noipa and co-workers [19] proposed to modify the method by using a cationic surfactant, which enabled analysis of the antioxidant activity of hydrophilic antioxidants contained in water extracts in the micelles formed in a solution.

3.1.2. *Trolox Equivalent Antioxidant Capacity (TEAC) assay*

An assay employing the $ABTS^{•+}$ cation-radical was proposed by Miller and co-workers [20]. It is based on a colour reaction, in which the stable cation-radical $ABTS^{•+}$ is formed from 2,2′-azinobis-(3-ethyl-benzothiazoline-6-sulfonic) acid (ABTS) with metmyoglobin and hydrogen peroxide. The reaction runs in phosphate-buffered saline, pH 7.4 (PBS). In a modification of the method proposed by Ozgen and co-workers [21], pH is equal to 4.5, which is to make it closer to that of the materials under analysis. A solution of the prepared radical turns blue-green, with the absorption spectrum within the range from approx 490 to 900 nm. When the antioxidants contained in the solution quench the $ABTS^{•+}$ cation-radical, the solution absorbance decreases, which is observed by colorimetry after 6 minutes of the reaction at the temperature of 30ºC and the wavelength of 734 nm. In the method modification proposed by Re and co-workers [22], the $ABTS^{•+}$ radical is generated in the reaction of 22′-azinobis-(3-ethyl-benzothiazoline-6-sulfonic acid) diammonium salt and potassium persulfate in dark at room temperature for 12-16 hours. The analysis results are expressed as an equivalent of the reference substance, e.g. vitamin C, gallic acid, and, most frequently, Trolox. Trolox, which is water-soluble vitamin E analogue, is used to plot the standard curve. Due to this, it is possible to express the strength of antioxidants under analysis in a unified scale TEAC and to compare the results achieved by different researchers.

3.1.3. *Ferric Ion Reducing Antioxidant Power (FRAP) assay*

Analysis of antioxidant activity by performing a FRAP assay was proposed by Benzie and Strain [23]. It involves colorimetric determination of the reaction mixture in which the oxidants contained in the sample reduce Fe^{3+} ions to Fe^{2+}. At low pH, Fe(III)-TPTZ (ferric-tripyridltriazine) complex is reduced to the ferrous (Fe^{2+}) form and intense blue colour at 593 nm can be observed. The FRAP reagent is prepared by mixing 2.5 ml of TPTZ (2,4,6-tris (1-pyridyl)-5-triazine) solution (10 mM in 40mM HCl), 25 ml acetate buffer, pH 3.6, and 2.5 ml $FeCl_3 \bullet H_2O$ (20 mM). The colour of $Fe(II)(TPTZ)_2$ which appears in the solution is measured colorimetrically after incubation at 37ºC. The measurement results are compared to those of a blank sample, which contains deionised water instead of the analysed sample. The duration of the assay differs from one study to another: 4 min [23, 24], 10 min [25] to 15 min [26]. The analysis results are converted and expressed with reference to a standard substance, which can be ascorbic acid [26], $FeSO_4$ [23, 25], Trolox [27,18].

3.1.4. *CUPric Reducing Antioxidant Capacity (CUPRAC) assay*

The CUPRAC assay, developed by Turkish researchers from Istanbul University [28], has undergone many modifications by which it has been adapted to wider applications [29, 30]. The mechanism of monitoring the antioxidant activity of the sample has remained unchanged. The assay is based on a coloured reaction during which copper ions in the CUPRAC reagent, Cu(II)-neocuproine (2,9-dimethyl-1,10-phenanthroline (Nc)), are reduced by antioxidants contained in the analysed sample. Chelates Cu(I)-Nc formed during the reaction have the maximum light absorption at the wavelength of 450 nm. The reaction runs at pH 7, which – as the authors of the method have pointed out – is closer to the natural physiological environment, unlike in the FRAP assay (pH 3.6) and the Folin-Ciocalteu assay (pH 10) [29, 30]. Those same authors have pointed out the low cost of the method, its simplicity, response to thiol groups of antioxidants and, importantly, its flexibility, which enables using it – by changing a solvent – to examine both lipophilic and hydrophilic antioxidants. The reaction of Cu(II)-Nc with an antioxidant runs vigorously at 37ºC, but certain compounds, such as naringine, require previous acidic hydrolysis.

3.1.5. DMPD (N,N-dimethyl-p-phenylenediamine) radical cation decolorization assay

A method of analysing antioxidant activity with respect to the $DMPD^{\bullet+}$ cation-radical (N,N-dimethyl-p-phenylenediamine cation radical) has been proposed by Fogliano and co-workers [31]. The determination principle involves colorimetric observation of the disappearance of the cation-radical colour at the absorbed light wavelength of 505 nm after a reaction time of 10 min. Coloured cation-radical $DMPD^{\bullet+}$ in the assay is obtained by reaction of DMPD with iron chloride in an acetate buffer at pH 5.25. The decrease in absorbance of the reaction mixture caused by antioxidants is compared to the calibration curve, prepared with a series of dilutions of Trolox [32].

Asghar and Khan [33] modified the method by adding $K_2S_2O_8$ (potassium persulfate) in an acetate buffer at pH 5.6 as the initiator of $DMPD^{\bullet+}$. They abandoned ferric chloride, due to the presence of metal ions in the analysed material which could – as a result of the Fenton Reaction – induce formation of hydroxyl radicals, which affects the antioxidant activity which is being determined. They also noted that the $DMPD^{\bullet+}$ radical obtained in the reaction with $K_2S_2O_8$ is more stable than that obtained with $FeCl_3$ as iron ions are susceptible to oxidation by atmospheric oxygen.

The improved $DPMD^{\bullet+}$ decolorization assay is suitable for water-soluble as well as lipid-soluble antioxidants [33]. A stock solution of DMPD cation radical is diluted to $A_{517.5nm}$ =0.7÷0.8 and after equilibration at 25ºC stabilized with ethanol or an acetate buffer (pH 5.6). The experiment is conducted at 30ºC and the absorbance of the reaction mixture is read out after 6 minutes. The measurement values obtained by the method with the cation radical DMPD are comparable with those obtained in the ABTS assay. As the cost of the DPMD is several times lower, it could be successfully used as an alternative for the ABTS assay [33].

3.1.6. Folin-Ciocalteu assay

The Falin-Ciocalteu reagent (FCR) is a complex formed in a reaction between sodium tungstate and sodium molybdenate in hydrochloric acid and phosphoric acid, which turns yellow after lithium sulphate is added. The reagent reacts in an alkaline environment with reducing compounds. Such a reaction gives a blue chromophore which is observed by colorimetry. The Folin-Ciocalteu method is highly sensitive – both to phenolic and non-phenolic compounds, e.g. proteins, vitamin C, vitamin B_1, folic acid, Cu(I). The method is applied most frequently to determine the total content of phenolic compounds [34, 35]. If that is the case, a sample for determination should be prepared in a proper manner to minimise the effect of non-phenolic compounds on the assay results. One such method is to remove the solvent from the sample and to dissolve phenolic compounds in alcohol, which eliminates the compounds insoluble in that environment or ones which become denatured.

Performing the assay is reduced to putting an alcoholic solution of the analysed sample, Folin-Ciocalteu reagent and solution of sodium carbonate into a reaction tube, which brings the pH of the reaction environment to approx. 10. According to various literature reports, the reaction runs in the darkness for 10 to 120 minutes. After that time, the blue colour of the solution is observed colorimetrically at 725 nm – 760 nm [34, 35, 36, 37, 38]. The results are expressed based on calibration curves prepared for catechol and gallic acid.

3.2. Hydrogen Atom Transfer (HAT) methods

3.2.1. Total Radical Trapping Antioxidant Parameter (TRAP) assay

This method is based on the measurement of the fluorescence of a molecular sample. Canadian researchers proposed this method to determine total peroxyl radical-trapping antioxidant capability of plasma [39]. They used a water-soluble azo compound, such as AAPH [2,2′-azo-bis-(2-amidinopropane)]. They measured the induction time electrochemically by measuring the time in which the antioxidants contained in the analysed sample prevent the capture of oxygen atoms. Four years later, DeLange and Glazer [40] proposed R-phycoerythrin in an induction measurement as a fluorescence indicator. They observed the time in which antioxidants in the sample protect R-phycoerythrin from oxidation and compared it with the time by which Trolox, added at a known amount, prevented a decrease in fluorescence. The reaction kinetics curve was monitored at the excitation wavelength of 495 and emission wavelength of 575, and antioxidant properties were expressed as Trolox equivalent. However, the method is time-consuming and rather complicated, which makes it susceptible to considerable errors, especially in inter-laboratory studies. Another modification of the method was developed by Valkonen and Kuusi [41], who used dichlorofluorescein diacetate (CDFH-DA) as an indicator of oxidation progress. DCFH-DA is hydrolysed in the presence of AAPH to highly fluorescent dichlorofluoresceine (DCF). An increase in fluorescence is a sign that the antioxidant activity of the analysed sample is exhausted.

3.2.2. Oxygen Radical Absorbance Capacity (ORAC) assay

The ORAC method was first proposed by Cao and co-workers in 1993. Like in the TRAP method, they used a fluorescent indicator. Determination of antioxidant activity by this method is based on measurement of decreasing fluorescence of the indicator caused by the radicals generated in the system. The reaction mixture in their proposal consisted of a fluorescent indicator β-phycoerythrin (β-PE), 2,2′-azobis(2-amidinopropane) dihydrochloride (AAPH) as a peroxyl radical generator and the analysed sample [42]. Attributing the low purity of β-phycoerythrin (approx. 30%) to the low reproducibility of fluorescence and the occurrence of different forms of phycoerythrin, Ou and co-workers [43] modified the method by replacing the indicator with fluorescein (3′,6′-dihydroxyspiro[isobenzofuran-1[3H],9′[9H]-xanthen]-3-one).

The reaction of antioxidants in a sample with the radicals generated by AAPH and fluorescein is conducted in a phosphate buffer at pH 7.4 and at the temperature of 37°C. As the reaction progresses, antioxidants in the analysed sample react with the radicals. With an excess of radicals, the ability of antioxidants to reduce them becomes exhausted and radicals react with fluorescein, oxidising it to a non-fluorescing form. Observation of the fluorescence of the reaction mixture is conducted at the excitation wavelength of 485 nm and emission wavelength of 525 nm. Measurements are conducted every 60-90 seconds until the resulting curve reaches a plateau.

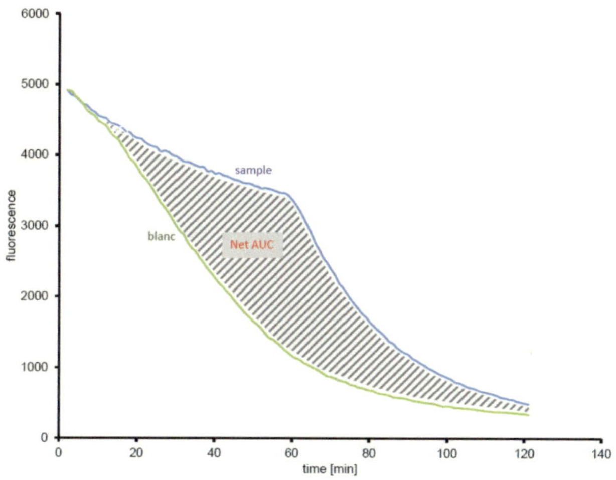

Figure 3. ORAC antioxidant activity determination of *Echium vulgare* defatted seeds methanolic extract expressed as net area under the curve (net AUC)

Surface area under the curve (AUC) was calculated by Ou and co-workers [43] from the following formula:

$$AUC = 1 + f_1 / f_o + f_2 / f_o + f_3 / f_o + f_4 / f_o + \ldots + f_{34} / f_o + f_{35} / f_o$$

where f_o denotes fluorescence read out at the beginning of the assay and f_i denoted the value of fluorescence read after the time i.

The area under curve for the sample (AUC_{sample}), reduced by the area under curve plotted for the blank sample (AUC_{blanc}) is referred to as "net AUC". Moreover, the net AUC is calculated for a series of dilutions of Trolox and a calibration curve is plotted, showing the relationship between net AUC and the concentration of Trolox. The results of the assay which refer the net AUC of the sample to the calibration curve are expressed as Trolox equivalent.

3.2.3. β-carotene bleaching test

Determination of the antioxidant activity in the system comprising β-carotene and linoleic acid is based on competitive oxidation of β-carotene during heat-induced auto-oxidation of linoleic acid. In the method proposed by Miller [44], a decrease in the absorbance of aqueous emulsion of "β-carotene – linoleic acid – analysed sample" depends on the antioxidant activity of the sample components. The antioxidant under study reacts with the radicals generated by linoleic acid in an incubated sample. As the ability of the analyte to scavenge radicals decreases, the oxidative effect on β-carotene increases. Measurement results of absorbance at 470 nm are read out every 15 minutes until the plateau is reached. The oxidative strength of the analyte is presented as the amount of β-carotene which was protected against oxidation.

3.2.4. Crocin Bleaching Assay (CBA)

Crocin is one of carotenoids present in saffron. It is present there as several isomers, differing by biological activity. The method of determination of antioxidant activity using crocin as an indicator was proposed in 1984 by Bors and co-workers [45]. In order to determine the antioxidant activity of the components of an analysed sample, it is put into a reaction tube together with solution of crocin diluted with phosphate buffer at pH 7.4. Thus obtained, the mixture is treated with radicals generated by solution of AAPH. The reaction runs at the temperature of 40°C. Decrease in the absorbance of the solution is measured colorimetrically at the wavelength of 443 nm and recorded for 10 minutes relative to the blank sample. The method has been modified many times [46, 47, 48]. Considering the problem of the unrepeatability of the composition of the saffron dye extract and, consequently, the differences in biological activity of the mixture of crocin isomers, Bathaie and co-workers [48] used α-crocin in their modification of the method. The results are expressed as "percent of inhibition of crocin degradation" (% Inh) and refer to the calibration curve prepared with Trolox and expressed as its equivalent (% Inh_{Trolox}).

3.3. Other methods

3.3.1. Cellular Antioxidant Activity (CAA) assay

A novel method of determination of antioxidant activity was proposed in 2007 by Wolfe and Liu of Cornell University [49]. They devised a method which is based on reactions running inside cells. According to them, the method better reflects reality than in-vitro methods due to the intake, metabolism and distribution of antioxidants in cells and, consequently, in a live organism. In a CAA assay, a solution of DCFH-DA (2',7'-Dichlorofluorescein diacetate) and a solution of the substances under analysis in PBS (phosphate buffered saline) at pH 7.4 is added to human hepatocarcinoma (HepG2) cells. Cells are incubated at 37°C and, during the incubation, DCFH-DA and components of the sample diffuse through the cell membrane into the cell. After that, the unabsorbed remainder is washed out with PBS and a solution of AAPH is added which – after its infiltration into a cell – generates free radicals oxidising DCFH-DA to DCF. Antioxidants quench the radicals, which reduces the amount of DCF, whose fluorescence is measured at 485 nm (ex.) and 520 nm (em.) during the analysis. The area under the fluorescence curve can be compared to the calibration curve prepared for a standard antioxidant, e.g. Trolox, and expressed as its equivalent. There are also modifications of the method, in which blood erythrocytes are used instead of HepG2 cells.

3.3.2. Total Oxyradical Scavenging Capacity (TOSCA) assay

Total Oxyradical Scavenging Capacity assay was proposed by Winston and co-workers as a rapid gas chromatographic method. They used this assay as a method of quantifiable measurement of the ability of sample antioxidants to quench free radicals [3]. The assay is based on the reaction between free radicals (peroxyl, hydroxyl, alkoxyl) and α-keto-γ-methiolbutyric acid (KMBA). The reaction yields ethylene, which can be simply analysed by gas chromatography. The assay involves incubation of solutions of AAPH, KMBA and the analysed sample at 39°C, with resulting ethylene production. Its content is determined every 12 minutes for the 96–120 minute period of the assay. The values obtained in the measurement form the basis for plotting the curve illustrating the changes in ethylene content. Quantitative determination of TOSC is possible only by comparison of the area under the curve for the analysed sample ($\int_{SA}$) and the control sample ($\int_{CA}$). The value of TOSC was calculated by Winston and co-workers as the difference between 100 and the ratio of the area for the analysed sample and the control sample, multiplied by 100.

$$TOSC = 100 - 100 * (\int_{SA} / \int_{CA})$$

When the radical inhibition reaches its theoretical maximum, ethylene is not produced and the value of TOSC is equal to 100 [3].

4. On-line liquid chromatography in bioactivity determination of compounds

The methods of determination of antioxidant activity are popular, which does not mean that they are only used in scientific research. A number of modifications of those methods along with methods which are not presented here are still used in analytical procedures applied in examination of bioactive substances. Owing to constantly broadening knowledge on the mechanism of oxidation and action of antioxidants, the choice and development of analytical methods is also changing. Increasing awareness of biological activity and the availability of analytical methods has changed the way substances are analysed. Regarding the different transport mechanisms of substances in organisms, observations have been conducted with different test cells. An analysis of different substances in mixtures has revealed differences in their biological activity. Due to such differences within a sample, it may contain both strong antioxidants and biologically inactive compounds as well as pro-oxidants. Their separation may obtain individual substances, or their mixtures, with beneficial biological properties. In search of rich sources of bioactive substances, screening studies are conducted in which isolated components of mixtures are analysed for their activity, e.g. as antioxidants. Time and money which must be spent on such analyses, as well as new testing capabilities, combined with chromatographic methods have made looking for such sources much cheaper and easier. Examples of using different chromatographic methods to inhibit or promote oxidation reactions are presented below.

4.1. TLC-methods for antioxidant activity analysis

Analysis of the biological activity of extracts by the methods presented above provides a researcher with a pooled result of the activities of all the components of a mixture. When analysing extracts of evening-primrose and starflower, Wettasinghe and Shahidi [50] made use of the experience gathered by Amarowicz and co-workers in fractionation of plant extracts [51, 52, 53] to achieve more precise characteristics of components of the extracts under analysis. They separated extracts by column chromatography with Sephadex LH20 column packing. As a result, they obtained six fractions, which they further analysed to determine their biological activity [50].

Separation of analyte fractions by column chromatography requires time, labour and money and the results show only properties of the properties of compounds in individual fractions. Using thin-layer chromatography in analysis of antioxidant activity of mixture components made it unnecessary to isolate them prior to analysis. Researchers from Kansas State University made use of the experience gathered by Marco [54] and Taga and co-workers [55] and proposed a method of determination of antioxidant activity of individual components of mixtures using the β-carotene bleaching assay for substances previously separated by TLC. They sprinkled β-carotene solution with linoleic acid on substances separated on a plate. They exposed the prepared plates to light and observed the disappearance of the orange colour of β-carotene. Spots with antioxidants were visible as ones with more intense colour because of their protective effect on β-carotene [56].

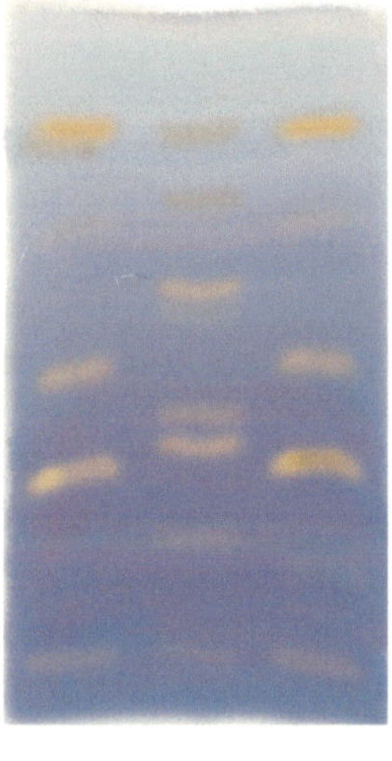

Figure 4. Compounds resolved on the TLC plate after spraying with DPPH methanolic solution.

Glavind and Holmer proposed a method of determination of antioxidants by TLC using the DPPH• radical. They sprinkled a plate with separated substances with methanol solution of the radical and observed discoloration where substances able to quench radicals were present [57]. The TLC-DPPH assay allows a researcher to access the analysed substances and to assess the biological activity of individual compounds. Another advantage of the method is the possibility of conducting screening analyses in which many extracts can be analysed. This enables effective and cheap searching for bioactive substances in unknown samples [58, 59, 60, 61].

4.2. The use of high performance liquid chromatography as a tool for bioactivity analysis

When seeking a tool which could be used to determine the biological activity of mixture components in a more precise way, researchers directed their attention towards high-performance liquid chromatography (HPLC). Its advantage over TLC is its higher resolution, which helps to avoid false results caused by the co-elution of different compounds.

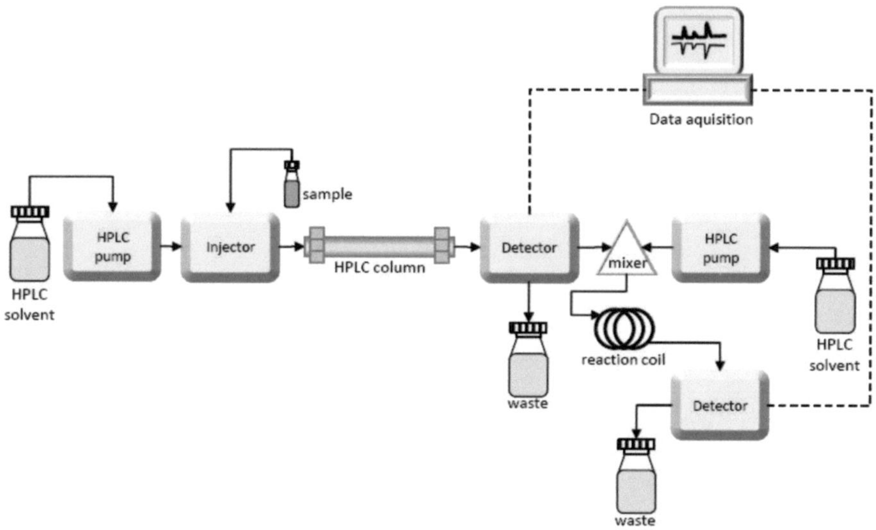

Figure 5. Instrumental setup for on-line HPLC radical scavenging assay

4.2.1. HPLC-DPPH• stable radicals decolorization

Initially, the use of HPLC in analysis of antioxidant properties with the DPPH• radical was restricted to chromatographic analysis of the radical content in solution. An assay was performed in which a solution of the radical was treated with the extract under analysis. The reaction ran in a reaction tube and the remainder of the radical after the reaction was analysed chromatographically. A comparison of the radical content in the blank sample and in the extract sample showed the amount of radical that was quenched by antioxidants in the analysed sample [62, 63]. However, the method did not provide more information than the colorimetric method. Much better results are obtained in a post-column on-line reaction in which substances separated on a chromatographic column react with a radical in a reaction coil.

A detector records a signal at 515 or 517 nm [64, 65]. Depending on the antioxidant activity of the separated substances, greater or smaller signal fading can be observed, resulting in a negative peak. The surface area of the peak, proportional to the antioxidant activity (compared to the standard curve plotted for Trolox) is the basis for expressing the result as its equivalent. An apparatus which can be used to conduct such an analysis should – like the basic HPLC set – consist of the main pump (feeding mobile phase to the system), an injecting device (injecting the sample), a column (which may be placed in a thermostat), a detector and a recording device. However, additional equipment must be used apart from the HPLC set. A solution of the DPPH$^{\bullet}$ is fed through an additional pump and, together with eluate, leaves the HPLC system to the mixer. Mixed substances are transferred to the reaction coil. A reaction coil, which is a capillary tube with a length ranging from 0.2 to 15 m, is where a reaction takes place between the mixture components and the DPPH reagent.

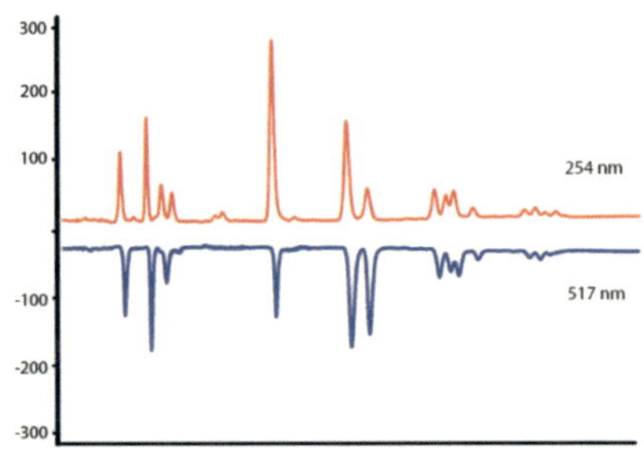

Figure 6. UV and DPPH radical quenching chromatogram of a plant methanolic extract.

Bartasiute and co-workers [66] analysed the method with capillaries of different lengths. The shortest capillary used in their experiment (0.2 m) ensured sufficient reaction time. However, specific properties of analysed mixtures should be taken into account, whose components may react differently. The last element is another detector which records the characteristic signal of the DPPH$^{\bullet}$ solution [65, 66, 67, 68]. Fading of a radical signal after reaction with an antioxidant (visible as a negative peak) is proportional to its oxidative force. Comparison of the surface area of the peak with the calibration curve prepared for Trolox shows the activities of the substances as equivalents of the antioxidant.

4.2.2. On-line HPLC-ABTS$^{\bullet+}$ stable radical decolorization

Using the ABTS$^{\bullet+}$ cation radical in the analysis of activity of mixture components in conjunction with HPLC was proposed in 2001 [69]. That method is similar to the HPLC-DPPH$^{\bullet}$ assay described above, but its sensitivity is higher. Components separated on an analytical column are transferred with a solution of the ABTS$^{\bullet+}$ cation radical to the reaction coil. A capillary with a length ranging from 1.5 to 15 metres is placed where the flow-through reaction of the radical quenching by the oxidants present in the sample takes place. The capillary length is selected depending on the expected reaction time. The signal characteristics of the ABTS$^{\bullet+}$ radical solution are recorded at 734 nm [69, 70] and in subsequent modifications – at 720 nm and 747 nm [16, 71, 72]. The antioxidant activity of individual compounds is determined based on the size of negative peaks which show the compound's ability to inactivate the radical. Comparison of the surface area of a negative peak caused by the presence of the analysed compound with the curve prepared for Trolox enables the activity of compounds to be expressed as the Trolox equivalent.

4.2.3. On-line HPLC-CUPRAC assay

An on-line HPLC-CUPRAC assay has been proposed by Çelik and co-workers. They used a chromatographic set with configuration used in on-line analysis of antioxidant properties with the $DPPH^{\bullet}$ and $ABTS^{\bullet +}$ radicals. Unlike the methods with the $DPPH^{\bullet}$ and $ABTS^{\bullet +}$ radicals, the assay is based on measurement of the growth of the solution colour intensity [73]. As in the off-line method, the Cu(II)-Nc reagent reacts with an antioxidant and is reduced to a yellow complex Cu(I)-Nc. The solution containing the complex has the absorbance with the maximum at 450 nm. The compounds separated chromatographically in the developed on-line method are mixed with the Cu(II)-Nc reagent in a mixer and are subsequently transferred to the reaction coil where a oxidation-reduction reaction takes place during its flow through the capillary. The antioxidant activity of individual compounds is observed as an increase in the signal on the detector at 450 nm. When a calibration curve is plotted with data obtained for Trolox, it is possible to express the compound's activity as an equivalent of Trolox activity [73].

4.2.4. On-line HPLC Crocin Bleaching Assay (HPLC-CBA)

The Crocin Bleaching Assay (CBA) in an off-line colorimetric version provided the base for developing the On-line HPLC-Crocin Bleaching Assay [74]. When a high-resolution method of compound separation is used, CBA can be used to determine the activities of each mixture component. The results are not affected by other mixture components, which improves the usability of the method for an objective assessment of bioactive components of mixtures. Like the off-line method described earlier, the authors of this method proposed crocin as the oxidation indicator and the AAPH reagent as the source of the radical. Antioxidants in a sample prevent oxidation of crocin by inactivating radicals generated by the AAPH reagent, as was the case in the colorimetric method. The signal recorded by the detector for 440 nm shows the antioxidant activity of each compound as chromatographic peaks with a surface area proportional to their activity. The mixture of crocin and the AAPH reagent was kept at 0ºC before being transferred to the system. The reaction mixture was combined on-line with eluate from the chromatographic column and the reaction between compounds ran during their flow through the reaction coil at 90ºC. The reaction parameters have a great effect on the interference caused by the detector; hence, the authors optimised the method, showing that the interference is affected by: instability of the reaction temperature, change of the AAPH:crocin ratio, the presence of air or nitrogen bubbles in the reaction coil and changes in the mobile phase composition [74]. Like other methods, it seems justified to express the results in a universal unit, i.e. the equivalent of a standard antioxidant, e.g. Trolox.

4.2.5. On-line chemiluminescence detection (HPLC-CL)

A sensitive on-line chemiluminescence method, "on-line HPLC-CL", was developed by Toyo'oka and co-workers [75]. This method helps to determine with high sensitivity the antioxidant activity of the separated compounds relative to H_2O_2 and $O_2^{-\bullet}$. In order to determine the activity of compounds relative to H_2O_2, a solution of luminol and H_2O_2 must be prepared. The activity of the analysed components relative to $O_2^{-\bullet}$ is measured in the system of reagents containing a mixture of luminol and hypoxanthine as well as a mixture of xanthine oxidase and catalase. The solutions, fed with two pumps, are mixed in a mixer before being joined with a stream of chromatographically-separated components which leave the column. Combining the streams of reagents and the analyte starts the reaction of radical quenching of the analysed sample by antioxidants. A decrease in the amount of radicals results in a decrease in the luminol luminescence intensity recorded by the detector. The compounds able to quench the radicals cause the signal to deviate from the base line, which is observed as negative peaks. The surface area of the peaks is proportional to the antioxidant activity of the analysed compounds. As is shown in the description, the method requires proper apparatus. The HPLC set, necessary to separate the mixture components under analysis has to be fitted out with two separate pumps and a mixer. The stream of eluate from the analytic column and the reagents are joined in a mixing module, after which the mixture is transferred to the reaction coil. When it flows through the reaction coil, radical capture reaction takes place and the other radicals react with luminol. The radiation generated in this way is recorded by a photomultiplier. A negative chromatogram generated in this manner is used as the basis for assessment of antioxidant activity of the separated compounds. Like the methods described above, the assessment is based on a comparison of surface area of peaks with the calibration curve prepared for a standard antioxidant, e.g. Trolox.

5. Conclusion

People have made use of the properties of different compounds without realising their existence for a long time. Since science found ways to determine the nature of the effects exerted by bioactive compounds (e.g. medicinal plants) analytical methods have been perfected to enable more detailed analysis of the material. Analyses have focused on determination of the intensity of biological activity and on identification of the components responsible for the activity. Many materials have been analysed in search of bio-components. Mastering chromatographic methods has provided the possibility of high-resolution analysis of compounds, including their biological activity. The analytical methods which have been characterised here are the result of several dozen years of research into improving analytical methods in the search for biologically active compounds. The proposed classification is based on the mechanism of reaction observed in assays. The methods make use of reactions induced by the presence of radicals generated as initiators of oxidation reactions prevented by analysed bioactive compounds. It is not the only possible approach to looking for and analysing bioactive compounds. Various methods of analysing antioxidant activity which have not been mentioned above have been applied on a marginal scale due to their drawbacks. Apart from that, there are methods of analysing biological activity which analyse substances capable of inhibition of/affinity to certain enzymes, e.g. acetylcholinesterase, phosphodiesterase, glutathione-S-transferase (EAD – Enzyme Activity/Affinity Detection), affinity of bioactive substances receptors, e.g. estrogen receptor (RAD – Receptor Affinity Detection) [76, 77, 78, 79, 80,

http://dx.doi.org/10.5772/55620

81, 82]. Conjunction of the chromatographic methods of component separation with methods of analysis of biological properties provides great opportunities in their analysis. This has made the search for bioactive substances easier and will aid the future development of new research methods.

Acknowledgements

This study was supported by the Grant N N312 466340 from the National Science Centre, Poland.

Author details

Sylwester Czaplicki

Chair of Food Plant Chemistry and Processing, Faculty of Food Sciences, University of Warmia and Mazury in Olsztyn, Olsztyn, Poland

References

[1] Witkiewicz Z, Kałużna-Czaplińska J. Basics of chromatography and electromigration techniques. Warsaw: WNT; 2012.

[2] Prior RL, Wu X, Schaich K. Standardized methods for the determination of antioxidant capacity and phenolics in foods and dietary supplements. Journal of Agricultural and Food Chemistry. 2005; 53(10) 4290–4302.

[3] Winston GW, Regoli F, Dugas AJ Jr, Fong JH, Blanchard KA. A rapid gas chromatographic assay for determining oxyradical scavenging capacity of antioxidants and biological fluids. Free Radic Biol Med. 1998; 24(3) 480-493.

[4] Brand-Williams W, Cuvelier ME, Berset C. Use of a free radical method to evaluate antioxidant activity. Food Science and Technology. 1995; 28 25-30.

[5] Ozcelik B, Lee JH, Min DB. Effects of Light, Oxygen, and pH on the Absorbance of 2,2-Diphenyl-1-picrylhydrazyl. Journal of Food Science. 2003; 68(2) 487–490.

[6] Bondet V, Brand-Williams W, Berset C. Kinetics and Mechanisms of Antioxidant Activity using the DPPH$^{\bullet}$ Free Radical Method. Food Science and Technology. 1997; 30, 609–615.

[7] Locatelli M, Gindro R, Travaglia F, Coïsson J-D, Rinaldi M, Arlorio M. Study of the DPPH-scavenging activity: Development of a free software for the correct interpretation of data. Food Chemistry. 2009; 114 889–897.

[8] Stankevičius, M.; Akuneca, I.; Jãkobsone, I.; Maruška, A. Analysis of phenolic compounds and radical scavenging activities of spice plants extracts. Food Chemistry and Technology. 2010; 44(2) 85-91.

[9] Moure A, Franco D, Sineiro J, Dominguez H, Nunez MJ, Lema JM. Antioxidant activity of extracts from Gevuina avellana and Rosa rubiginosa defatted seeds. Food Research International. 2001; 34(2/3) 103-109.

[10] Hatano T, Kagawa H, Yasuhara T, Okuda T. Two new flavonoids and other constituents in licorice root: their relative astringency and radical scavenging effects. Chemical & Pharmaceulical Bulletin. 1988; 36(6) 2090-2097.

[11] Müller L, Fröhlich K, Böhm V. Comparative antioxidant activities of carotenoids measured by ferric reducing antioxidant power (FRAP), ABTS bleaching assay (αTEAC), DPPH assay and peroxyl radical scavenging assay. Food Chemistry. 2011; 129 139–148

[12] Marxen K, Vanselow KH, Lippemeier S, Hintze R, Ruser A, Hansen U-P. Determination of DPPH Radical Oxidation Caused by Methanolic Extracts of Some Microalgal Species by Linear Regression Analysis of Spectrophotometric Measurements. Sensors. 2007; 7, 2080-2095.

[13] Mandal P, Misra TK, Ghosal M. Free-radical scavenging activity and phytochemical analysis in the leaf and stem of *Drymaria diandra* Blume. International Journal of Integrative Biology. 2009; 7(2) 80-84.

[14] Wojdyło A, Oszmiański J, Czemerys R. Antioxidant activity and phenolic compounds in 32 selected herbs. Food Chemistry. 2007; 105 940–949.

[15] Wu JH, Huang CY, Tung YT, Chang ST. Online RP-HPLC-DPPH screening method for detection of radical-scavenging phytochemicals from flowers of Acacia confusa. Journal of Agricultural and Food Chemistry. 2008; 56(2) 328-332.

[16] Gong Y, Liu X, He W-H, Xu H-G, Yuan F, Gao Y-X. Investigation into the antioxidant activity and chemical composition of alcoholic extracts from defatted marigold (*Tagetes erecta* L.) residue. Fitoterapia 2012; 83 481–489.

[17] Sánchez-Moreno C, Larrauri JA, Saura-Calixto F. A Procedure to Measure the Antiradical Efficiency of Polyphenols. Journal of the Science of Food and Agriculture. 1998; 76 270-276.

[18] Michel T, Destandau E, GLe Floch G, Lucchesi ME, Elfakir C. Antimicrobial, antioxidant and phytochemical investigations of sea buckthorn (Hippophaë rhamnoides L.) leaf, stem, root and seed. Food Chemistry. 2012; 131 754–760.

[19] Noipa T, Srijaranai S, Tuntulani T, Ngeontae W. New approach for evaluation of the antioxidant capacity based on scavenging DPPH free radical in micelle systems. Food Research International. 2011; 44(3) 798–806.

[20] Miller NJ, Rice-Evans C, Davies MJ, Gopinathan V, Milner A. A novel method for measuring antioxidant capacity and its application to monitoring the antioxidant status in premature neonates. Clinical Science. 1993; 84 407-412.

[21] Ozgen M, Reese RN, Tulio AZ Jr, Scheerens JC, Miller AR. Modified 2,2-azino-bis-3-ethylbenzothiazoline-6-sulfonic acid (abts) method to measure antioxidant capacity of Selected small fruits and comparison to ferric reducing antioxidant power (FRAP) and 2,2'-diphenyl-1-picrylhydrazyl (DPPH) methods. Journal of Agricultural and Food Chemistry. 2006; 54(4) 1151-1157.

[22] Re R, Pellegrini N, Proteggente A, Pannala A, Yang M, Rice-Evans C. Antioxidant activity applying an improved ABTS radical cation decolorization assay. Free Radical Biology & Medicine, 1999; 26(9/10) 1231–1237.

[23] Benzie IFF, Strain JJ. The Ferric Reducing Ability of Plasma (FRAP) as a Measure of ''Antioxidant Power'': The FRAP Assay. Analytical Biochemistry. 1996; 239 70–76.

[24] Panda N, Kaur H, Mohanty TK. Reproductive performance of dairy buffaloes supplemented with varying levels of vitamin E. Asian Australasian Journal of Animal Sciences. 2006; 19(1) 19-25.

[25] Oveisi MR, Sadeghi N, Jannat B, Hajimahmoodi M, Behfar A, Jannat F, Nasaba FM. Human Breast Milk Provides Better Antioxidant Capacity than Infant Formula. Iranian Journal of Pharmaceutical Research. 2010; 9(4) 445-449.

[26] Valvi SR, Rathod VS, Yesane DP. Screening of three wild edible fruits for their antioxidant potential. Current Botany. 2011; 2(1) 48-52.

[27] Thaipong K, Boonprakob U, Crosby K, Cisneros-Zevallos L, Byrne DH. Comparison of ABTS, DPPH, FRAP, and ORAC assays for estimating antioxidant activity from guava fruit extracts. Journal of Food Composition and Analysis. 2006; 19 669–675.

[28] Apak R, Güçlü K, Özyürek M, Karademir SE. Novel Total Antioxidant Capacity Index for Dietary Polyphenols and Vitamins C and E, Using Their Cupric Ion Reducing Capability in the Presence of Neocuproine: CUPRAC Method. Journal of Agricultural and Food Chemistry. 2004; 52 (26) 7970–7981.

[29] Çelik SE, Özyürek M, Güçlü K, Apak R. CUPRAC total antioxidant capacity assay of lipophilic antioxidants in combination with hydrophilic antioxidants using the macrocyclic oligosaccharide methyl b-cyclodextrin as the solubility enhancer. Reactive & Functional Polymers. 2007; 67 1548–1560.

[30] Çelik SE, Özyürek M, Güçlü K, Apak R. Solvent effects on the antioxidant capacity of lipophilic and hydrophilic antioxidants measured by CUPRAC, ABTS/persulphate and FRAP methods. Tantala 2010; 81 1300–1309.

[31] Fogliano V, Verde V, Randazzo G, Ritieni A. Method for measuring antioxidant activity and its application to monitoring the antioxidant capacity of wines. Journal of Agricultural and Food Chemistry. 1999; 47(3) 1035-40.

[32] Busuricu F, Negranu-Pârjol T, Balaban DP, Popescu A, Anghel A. The evaluation of the wines antioxidant activity. Innovative Romanian Food Biotechnology. 2008; 2(2) 10-18.

[33] Asghar MN, Khan IU, Arshad MN, Sherin L. Evaluation of antioxidant activity using an improved dmpd radical cation decolorization assay. Acta Chimica Slovenica. 2007; 54(2) 295–300.

[34] Huang D, Ou B, Prior RL. The chemistry behind antioxidant capacity assays. Journal of Agricultural and Food Chemistry. 2005; 53(6) 1841-1856.

[35] Agbor GA, Oben JE, Ngogang JY, Xinxing C, Vinson JA. Antioxidant Capacity of Some Herbs/Spices from Cameroon: A Comparative Study of Two Methods. Journal of Agricultural and Food Chemistry. 2005; 53(17) 6819-6824.

[36] Krishnaiah D, Sarbatly R, Nithyanandam R. A review of the antioxidant potential of medicinal plant species. Food and Bioproducts Processing. 2011; 8 9 217–233.

[37] Abe LT, Lajolo FM, Genovese MI. Comparison of phenol content and antioxidant capacity of nuts. Ciência e Tecnologia de Alimentos 2010; 30(1) 254-259.

[38] Habib R, Rahman M, Mannan A, Zulfiker AH, Uddin ME, Sayeed MA. Evaluation of antioxidant, cytotoxic, antibacterial potential and phytochemical screening of chloroform extract of *Phyllanthus acidus*. International Journal of Applied Biology and Pharmaceutical Technology. 2011; 2(1) 420-427.

[39] Wayner DD, Burton GW, Ingold KU, Locke S. Quantitative measurement of the total, peroxyl radical-trapping antioxidant capability of human blood plasma by controlled peroxidation. The important contribution made by plasma proteins. Federation of European Biochemical Societies Letters 1985; 187(1) 33-37.

[40] DeLange RJ, Glazer AN. Phycoerythrin fluorescence-based assay for peroxy radicals: a screen for biologically relevant protective agents. Analytical Biochemistry. 1989; 177(2) 300-306.

[41] Valkonen M, Kuusi T. Spectrophotometric assay for total peroxyl radical-trapping antioxidant potential in human serum. Journal of Lipid Research. 1997; 38(4) 823-33.

[42] Cao G, Alessio HM, Cutler RG. Oxygen-radical absorbance capacity assay for antioxidants. Free Radical Biology & Medicine. 1993; 14(3) 303-11.

[43] Ou B, Hampsch-Woodill M, Prior RL. Development and Validation of an Improved Oxygen Radical Absorbance Capacity Assay Using Fluorescein as the Fluorescent Probe. Journal of Agricultural and Food Chemistry. 2001; 49 4619-4626.

[44] Miller HE. A simplified method for the evaluation of antioxidants, Journal Of The American Oil Chemists' Society. 1971; 48(2) 91.

[45] Bors W, Michel C, Saran M. Inhibition of the bleaching of the carotenoid crocin a rapid test for quantifying antioxidant activity. Biochimica et Biophysica Acta (BBA)/ Lipids and Lipid Metabolism. 1984; 796(3) 312-319.

[46] Tubaro F, Micossi E, Ursini F. The antioxidant capacity of complex mixtures by kinetic analysis of crocin bleaching inhibition. Journal of the American Oil Chemists' Society. 1996; 73(2) 173-179.

[47] Ordoudi SA, Tsimidou MZ. Crocin bleaching assay (CBA) in structure-radical scavenging activity studies of selected phenolic compounds. Journal of Agricultural and Food Chemistry. 2006; 54(25) 9347-9356.

[48] Bathaie SZ, Kermani FMZ, Shams A. Crocin Bleaching Assay Using Purified Di-gentiobiosyl Crocin (α-crocin) from Iranian Saffron. Iranian Journal of Basic Medical Sciences. 2011; 14(5) 399-406.

[49] Wolfe KL, Liu RH. Cellular Antioxidant Activity (CAA) Assay for Assessing Antioxidants, Foods, and Dietary Supplements. Journal of Agricultural and Food Chemistry. 2007; 55(22) 8896-8907.

[50] Wettasinghe M, Shahidi F. Scavenging of reactive-oxygen species and DPPH free radicals by extracts of borage and evening primrose meals. Food Chemistry. 2000; 70 17-26.

[51] Amarowicz R, Koslowska H, Shimoyamada M. Chromatographic analysis of rapseed glucosides fractions. Polish Journal of Food and Nutrition. 1992; 1 89-93.

[52] Amarowicz R, Wanasundara U, Wanasundara J, Shahidi F. Antioxidant activity of ethanolic extracts of faxseed in a β-carotene-linoleate model system. Journal of Food Lipids. 1993; 1 111-117.

[53] Amarowicz R, Shahidi F. A rapid chromatographic method for separation of individual catechins from green tea. Food Research International 1996; 29 71-76.

[54] Marco JG. A rapid method for evaluation of antioxidants. Journal Of The American Oil Chemists' Society. 1968; 45(9) 594-598.

[55] Taga SM, Miller EE, Pratt DE. Chia Seeds as a Source of Natural Lipid Antioxidants. Journal of the American Oil Chemists' Society. 1984; 61(5) 928-931.

[56] Mehta RL, Zayas JF, Yang S-S. Ajowan as a Source of Natural Lipid Antioxidant. Journal of Agricultural and Food Chemistry. 1994; 42(7) 1420–1422.

[57] Glavind J, Holmer G. Thin-layer chromatographic determination of antioxidants by the stable free radical α, α'-diphenyl-β-picrylhydrazyl. Journal of the American Oil Chemists' Society. 1967; 44(9) 539 – 542.

[58] Masoko P, Eloffa JN. Screening of Twenty-Four South African Combretum and Six Terminalia Species (Combretaceae) for Antioxidant Activities. African Journal of Traditional, Complementary and Alternative medicines. 2007; 4(2) 231–239.

[59] Jothy SL, Zuraini Z, Sasidharan S. Phytochemicals screening, DPPH free radical scavenging and xanthine oxidase inhibitiory activities of Cassia fistula seeds extract. Journal of Medicinal Plants Research. 2011; 5(10) 1941-1947.

[60] Cieśla Ł, Kryszeń J, Stochmal A, Oleszek W, Waksmundzka-Hajnos M. Approach to develop a standardized TLC-DPPH• test for assessing free radical scavenging properties of selected phenolic compounds. Journal of Pharmaceutical and Biomedical Analysis. 2012; 70 126– 135.

[61] Olech M, Komsta Ł, Nowak R, Cieśla Ł, Waksmundzka-Hajnos M. Investigation of antiradical activity of plant material by thin-layer chromatography with image processing. Food Chemistry. 2012; 132 549–553.

[62] Yamaguchi T, Takamura H, Matoba T, Terao J. HPLC method for evaluation of the free radical-scavenging activity of foods by using 1,1-diphenyl-2-picrylhydrazyl. Bioscience, Biotechnology, and Biochemistry. 1998; 62(6) 1201-1204.

[63] Bhandari P, Kumar N, Singh B, Ahuja PS. Online HPLC-DPPH method for antioxidant activity of *Picrorhiza kurroa* Royle ex Benth. and characterization of kutkoside by Ultra-Performance LC-electrospray ionization quadrupole time-of-flight mass spectrometry. Indian Journal of Experimental Biology. 2010; 48 323-328.

[64] Bandoniene D, Murkovic M. On-Line HPLC-DPPH Screening Method for Evaluation of Radical Scavenging Phenols Extracted from Apples (Malus domestica L.). Journal of Agricultural and Food Chemistry. 2002; 50, 2482-2487.

[65] Koleva II, Niederländer HA, van Been TA. An on-line HPLC method for detection of radical scavenging compounds in complex mixtures. Analytical Chemistry. 2000; 72(10) 2323-2328.

[66] Bartasiute A, Westerink BHC, Verpoorte E, Niederländer HAG. Improving the in vivo predictability of an on-line HPLC stable free radical decoloration assay for antioxidant activity in methanol–buffer medium. Free Radical Biology & Medicine. 2007; 42 413–423.

[67] Dapkevicius A, van Beek TA, Niederländer HAG. Evaluation and comparison of two improved techniques for the on-line detection of antioxidants in liquid chromatography eluates. Journal of Chromatography A. 2001; 912 73–82.

[68] Zhang Q, van der Klift EJ, Janssen HG, van Beek TA. An on-line normal-phase high performance liquid chromatography method for the rapid detection of radical scavengers in non-polar food matrixes. Journal of Chromatography A. 2009; 1216 7268–7274.

[69] Koleva II, Niederländer HA, van Been TA. Application of ABTS radical cation for selective on-line detection of radical scavengers in HPLC eluates. Analytical Chemistry. 2001; 73(14) 3373–3381.

[70] Yildiz L, Başkan KS, Tütem E, Apak R. Combined HPLC-CUPRAC (cupric ion reducing antioxidant capacity) assay of parsley, celery leaves, and nettle. Talanta. 2008; 77(1) 304-313.

[71] Stalmach A, Mullen W, Nagai C, Crozier A. On-line HPLC analysis of the antioxidant activity of phenolic compounds in brewed, paper-filtered coffee. Brazilian Journal of Plant Physiology. 2006; 18(1) 253-262.

[72] Stewart AJ, Mullen W, Crozier A. On-line high-performance liquid chromatography analysis of the antioxidant activity of phenolic compounds in green and black tea. Molecular Nutrition & Food Research. 2005; 49(1) 52–60.

[73] Çelik SE, Özyürek M, Güçlü K, Apak R. Determination of antioxidants by a novel on-line HPLC-cupric reducing antioxidant capacity (CUPRAC) assay with post-column detection. Analytica Chimica Acta. 2010; 674 79–88.

[74] Bountagkidoua O, van der Klift EJC, Tsimidou MZ, Ordoudi SA, van Beek TA. An on-line high performance liquid chromatography-crocin bleaching assay for detection of antioxidants. Journal of Chromatography A. 2012; 1237 80– 85.

[75] Toyo'oka T, Kashiwazaki T, Kato M. On-line screening methods for antioxidants scavenging superoxide anion radical and hydrogen peroxide by liquid chromatography with indirect chemiluminescence detection. Talanta. 2003; 60 467-475.

[76] Ingkaninan K, de Best CM, van der Heijden R, Hofte AJ, Karabatak B, Irth H, Tjaden UR, van der Greef J, Verpoorte R. High-performance liquid chromatography with on-line coupled UV, mass spectrometric and biochemical detection for identification of acetylcholinesterase inhibitors from natural products. Journal of Chromatography A. 2000; 872(1-2) 61-73.

[77] Andrisano V, Bartolini M, Gotti R, Cavrini V, Felix G. Determination of inhibitors' potency (IC_{50}) by a direct high-performance liquid chromatographic method on an immobilised acetylcholinesterase column. Journal of Chromatography B. 2001; 753 375–383.

[78] Schobel U, Frenay M, van Elswijk DA, McAndrews JM, Long KR, Olson LM, Bobzin SC, Irth H. High resolution screening of plant natural product extracts for estrogen receptor alpha and beta binding activity using an online HPLC-MS biochemical detection system. Journal of Biomolecular Screening. 2001; 6(5) 291-303.

[79] van Liempd SM, Kool J, Niessen WM, van Elswijk DE, Irth H, Vermeulen NP. On-line formation, separation, and estrogen receptor affinity screening of cytochrome P450-derived metabolites of selective estrogen receptor modulators. Drug Metabolism and Disposition. 2006;34(9) 1640-1649.

[80] Alphonse MP, Saffar AS, Shan L, HayGlass KT, Simons FER, Gounni AF. Regulation of the High Affinity IgE Receptor (FceRI) in Human Neutrophils: Role of Seasonal Allergen Exposure and Th-2 Cytokines. PLoS ONE. 2008; 3(4) e1921. doi:10.1371/journal.pone.0001921.

[81] Mroczek T. Highly efficient, selective and sensitive molecular screening of acetylcholinesterase inhibitors of natural origin by solid-phase extraction-liquid chromatography/electrospray ionisation-octopole-orthogonal acceleration time-of-flight-mass spectrometry and novel thin-layer chromatography-based bioautography. Journal of Chromatography A. 2009; 1216 2519–2528.

[82] Shi S-Y, Zhou H-H, Zhang Y-P, Jiang X-Y, Chen X-Q, Huang K-L. Coupling HPLC to on-line, post-column (bio)chemical assays for high-resolution screening of bioactive compounds from complex mixtures. Trends in Analytical Chemistry. 2009; 28(7) 865-877.

Chapter 5

Chromatographic Separations with Selected Supported Chelating Agents

Dean F. Martin

Additional information is available at the end of the chapter

http://dx.doi.org/10.5772/55521

1. Introduction

In this chapter, *ligands* have a more focused definition than they typically do. Ligands are molecules having donor atoms with a pair of electrons that form coordinate covalent bonds with metal ions. Unidentate ligands have a single donor atom, while chelating agents have two or more donor atoms, and in attaching to the metal ions form rings that are associated with enhanced stability. Further, two kinds of chelating agents are involved: *molecular* and *supported.*

Molecular chelating agents, such as ethylenediaminetetraacetic acid, $(HOOCCH_2)_2NCH_2$ - $CH_2N(CH_2COOH)_2$ EDTA react with metal ions, such as magnesium, calcium, or transition metal ions to form soluble molecular entities in aqueous solutions. These chelating agents are said to "sequester" the metal ions, much as a prisoner in jail is sequestered or removed from society. The metal ion is sequestered in the sense that the act of forming the compound changes the properties of the metal ion. For example, EDTA may be added in small quantities to beer to prevent a haze in the solution due to formation of insoluble, dispersed calcium carbonate. Two molecules of iminodiacetic acid, HN($CH_2N(CH_2COOH)_2$, H_2L may react to form soluble complexes of the type$Mg(L)_2{}^{=}$. Other ligand-complex entities may be insoluble, but the coordination compounds formed, soluble or insoluble, are molecular entities.

1.1. Examples of supported chelating agents

Supported chelating agents are attached physically or chemically to a solid, and the coordination entities that result from reaction of metal ions are insoluble polymeric materials. Iminodiacetic acid can be attached to a solid and is one of six chelating agents attached to a solid and sold as "QuadriPure™ Scavengers" sold by Sigma/Aldrich. One of these may be depicted as Solid—$CH_2N(CH_2N(CH_2COOH)_2$.

"Chelex 100" is another example of a commercially available supported chelating agent. The material was available from Bio-Rad and was used to purify compounds, most notably transition metal ions. [1] The preference for transition metal ions, e.g., copper (II) or iron(II) ions, over such univalent metal ions as sodium or potassium is said to be about 5000 to 1 [1]. The material consists of a styrene-divinylbenzene copolymer to which is attached iminodiacetic acid moieties.

In addition, Chelex 100 can be used for purification of DNA [2]. Though the mechanism of action seems uncertain, a plausible suggestion is that Chelex protects DNA from the effect of heating that is used to release DNA from cells. In addition, Chelex can sequester magnesium and heavy metal ions that would adversely affect the DNA. After treatment, DNA and RNA remain in the aqueous supernatant above the Chelex sample.

Supported chelating agents serve three major functions, i.e., concentration, elimination, and recycling metal ions.

Concentration of metal ions is a significant function that enhances analytical capabilities. Measuring the concentration of trace metals in dilute sea water is a significant analytical challenge. Using supported chelators allows an analyst to concentrate transition metals present in trace amounts from a large volume of sea water onto a solid in, say, a chromatography column, then eluting with dilute nitric acid to obtain a suitably concentrated solution. The technique can also be used to eliminate matrix effects. Unfortunately, Chelex seems to be less effective in sea water than in fresh water [3].

Also, users of Chelex may encounter problems with certain natural waters that are colored with samples of naturally occurring chelating agents such as soil organic acids (e.g., fulvic or humic acids) that provide competition for metal ions and reduce the effectiveness of Chelex [4,5]. Ryan and Weber [6] studied this problem and made pre-concentration comparisons of trace levels of copper ion in standard solutions comparing Chelex-100 with three different chelating agents supported on silica. The reagents tested were N-propylethylenediamine, the bis-dithiobicarbonate of that compound, and 8-hydroxyquinoline. The results indicated that in the presence of organic compounds typical organic-rich natural waters reduced the effectiveness of Chelex-100 in comparison with the three silica-supported agents. In general, immobilized 8-quinolinol was the most effective, as prepared by a modified Hill procedure [7].It is not the purpose of this review to criticize use of a successful commercial supported chelator; rather one may note that no such product can be expected to be highly effective under all conditions and circumstances, so it seems appropriate to consider examples of custom-made agents.

Elimination of metal ions is a second major function of supported chelators. And a variety of conditions may be envisioned. Electroplating firms can be faced with the need to remove metal ions from solutions as part of an EPA-approved disposal protocol. A radiator shop may be faced with a need to remove waste zinc ion before allowing water to go into a sewer, in an effort to minimize the danger of killing bacteria used in sewage treatment procedures. The Berkeley Pit (*vide infra*) represents an outstanding example of the need to remove waste copper.

Elimination of metals by chelators benefits from a recognition that the ability of the chelator depends on several factors, the size of the metal-chelate ring, the donor atoms involved, and

http://dx.doi.org/10.5772/55521

the metal ions involved. For example, the chelating tendency, expressed as log K, where K is the equilibrium constant for the formation of the metal complex, follows a certain order for many chelating agents. That order for divalent metal ions of the first transition metal series is known as the Irving-Williams Order [8] and may be represented as

Sc< Ti< Cr< Mn <Fe< Co< Ni<Cu >Zn

This may be compared with the results reported for a supported material called Octolig® (Table 1)

Divalent metal	Hg	Fe	Co	Ni	Cu	Zn
Log K	29	11.1	15.6	19.1	22.4	16.2

Table 1. Calculated formation constants for Octolig® [9, 10]

Recycling metal ions is part of the concentration process noted above. It can typically be achieved by treatment of the sorbed metals with dilute nitric acid. This is a strong acid, and the nitrate ion is not a strongly coordinating anion. In addition, recycling can be a useful means of paying for the removal process. An example of this was a laboratory bench-scale experiment to treat water from the Berkley Pit, recycle the copper and sell the copper to help pay the processing costs. The Berkley Pit [11] was once an open pit copper mine that was opened in 1955 and closed in 1982.

When the mine was closed, water pumps in a 3,800 ft shaft were turned off and the pit was slowly filled [11]. It is now a mile long, a half mile wide, and about 900 feet deep. It is said to be one of the largest Superfund sites. The chemistry of the Pit is complicated, one may say unfavorable, and certainly challenging. The bench-scale project considered the possibility of using Octolig® to remove copper and help pay for the process, but additional study would be needed [12].

2. Preparation of supported chelating agents

These materials can be made in several ways depending upon the method of linkage that is chosen. The procedures can be placed in three categories. They are covalent linkage, ionic linkage, and London-forces linkages. As an example of the covalent linkage, it is interesting to compare (*vide infra*) two sources: the detail in the patent (Examples 1-4) [14] with a journal article by Gao and co- workers [13], as in the experiments section 2.3 (entitled "Preparing and characterizing of composite absorption particles of PEI/SiO_2 ").

2.1. Covalent linkages

Lindoy and Eaglen [14] noted that metal ions were removed by the material produced and noted in Figure 1. A special feature was that due to the spacing established by the three-carbon bridge or linkage " the material maintains a high ion-complexation capacity." These workers also demonstrated metal-ion removal capacity for representative metals (transition metal ions, alkali metal ions, etc). The material under dynamic conditions removed 50 mg Cu(II) /g composite. The metal absorbing ability followed the order $Cu^{2+} > Cd^{2+} > Zn^{2+}$.

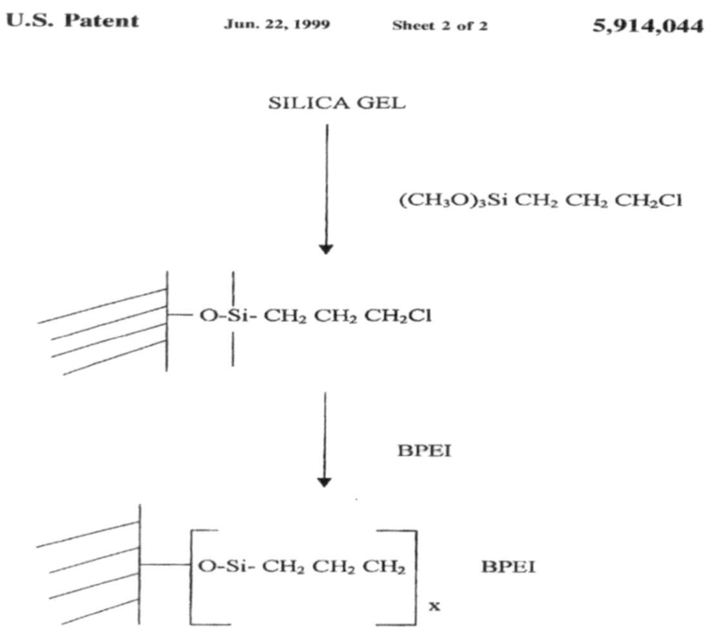

Figure 1. Flow chart for the synthesis of a supported chelating agent. BPEI consists of branched ethylenediimino moieties [14]

A similar technique was used by Soliman, who also used a silica gel matrix and a covalent linker to tie to a series of amines, mono-, di-, tri-, and tetra-amine [15]. Using a batch equilibrium technique, he measured the removal capacities (mmole/g) for divalent forms of cobalt, nickel, copper, zinc, cadmium, and lead. In general maximum removal values (at optimum pH values) were obtained for the tetra-amine species.

El Ashgar used a variation on the technique [16]. Specifically, 3-chloropropyltrimethoxysilane was used to alkylate diethylenetriamine, i.e, reaction with the halide end to produce a precursor of a polymer (I, Eqn. 1). The reaction of I with tetraethylorthosilicate resulted in a "diethylenetriamine polysiloxane immobilized ligand system."

$$\underset{\text{I}}{(CH_3O)_3Si(CH_2)_3NH(CH_2)_2NH(CH_2)_2NH_2} + 2\,(EtO)_4Si \rightarrow \text{Polymer} \quad (1)$$

The system was effective as a solid-phase preconcentration agent for cobalt(II), nickel(II), and copper(II) at pH 5.5 using column chromatography. A especially desirable feature was that the three metals could be cleanly separated by adjusting the pH of the eluent [16].

Another covalent approach is condensation of a potential ligand of the type HO-Y- Z where Y is a hydrocarbon moiety and Z is a coordinating group NH or NH_2 or SH. The mixture of solid substrate (clay, silica) was suspended in toluene with the potential ligands and a drop of sulfuric acid in a flask attached to a Dean-Stark tube, which is attached to a condenser. The reaction occurs between OH on the potential ligand and OH group on silica with the elimination of water using azeotropic distillation. Water produced in the reaction flask (2, Figure 2) distills as a constant-boiling mixture (3), that is cooled in a condenser (5), and separates into two components, which drop into a calibrated Dean-Stark tube (8). The water drops to the bottom of the tube; toluene returns to the reaction flask (2). The progress of the reaction can be measured, e.g., condensation of 0.1 mole of HO-Y-Z should give 1.8 milliliters of water. The reaction is allowed to proceed until completion [17, 20]. Clays were also used as supports [18, 19].

2.2. Ionic linkages

Examples are available from a number of studies. The authors demonstrated that the use of ion-exchange resins as supports for chelators that can be derivatized or converted to ions is a versatile technique [22]. Examples involving two different ion-exchange resins help indicate the range of the technique.

An *anion-exchange resin* in the hydroxide form RNHOH (e.g., IR-120), can be treated with other chelating agents HCh converted to the anionic forms, NaCh (Eqn. 2).

$$RNH_4OH + NaCh \rightarrow RNH_4Ch + Na^+ + OH^- \qquad (2)$$

A *cation-exchange resin* exists as a polyalkylsulfonic acid, RSO_3H and can react with a chelating agent in a protonated form HCh^+ (Eqn. 3).

$$RSO_3H + HCh^+ \rightarrow RSO_3H\,Ch + H+ \qquad (3)$$

Lee and coworkers [23] used the technique shown (Eqn 2) to load a number of chelating agents, among them chromotropic acid, onto the anion exchange resin Dowex® 1-X8 (chloride form). These composites are easy to prepare, and loading of metal ions on the chromatography columns showed the metal ion loading as a color change.

An anion exchange resin (e.g. Amberlite® R-120) was treated with protonated dithiooxamine, $H_2NC(S)C(S)NH_2$. Using the supported ligand, quantitative removal of copper, cadmium, and lead ion solutions at neutral or slightly alkaline solutions of deionized or tap water, but poor results were obtained with sea water [22].

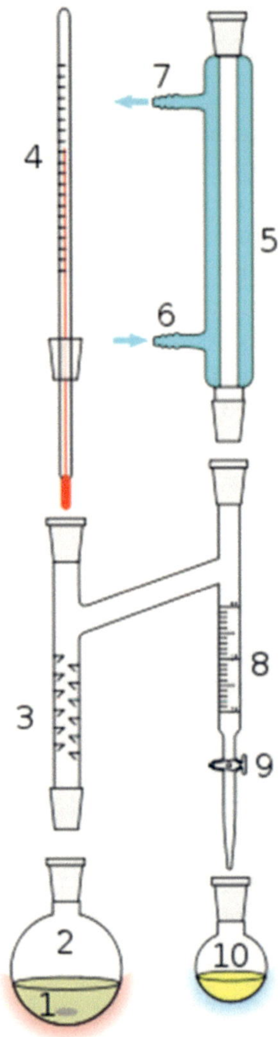

Figure 2. Schematic representation of a Dean-Stark Apparatus [21]

2.3. London-forces (corkscrew) approach

On occasion useful ligands are available, but need to be converted to supported ligands.

Examples of such ligands include LIX54 [$CH_3(CH_2)_{11}C_6H_4COCH_2COCH_3$], LIX860 (*n*-dodecylsaldoximine), oleoylacetone, and *N, N'*-didodecyldithiooxamide (all were once commercially available). A two-step procedure was used [24-27]: First, silica gel (5 g) was mixed with 25 mL of hexane, and the excess solvent allowed to evaporate to produce silica gel with pores loaded with hexane. Next the treated silica gel was mixed with a solution of a chelating agent in hot hexane. After stirring, the hexane was allowed to evaporate. It was hypothesized that a change in environment resulted with the evaporation of the hexane, i.e., that the long-chain hydrocarbon in the pores experienced changed environment, e.g., from hydrophobic to a hydrophilic condition. It was presumed that in the new environment, the form of the long- chain hydrocarbon altered, going from a stretched- out to a coiled version (idealized form, Figure 3) as a result of association of London forces. The coiled version would lead to a wedged ligand.

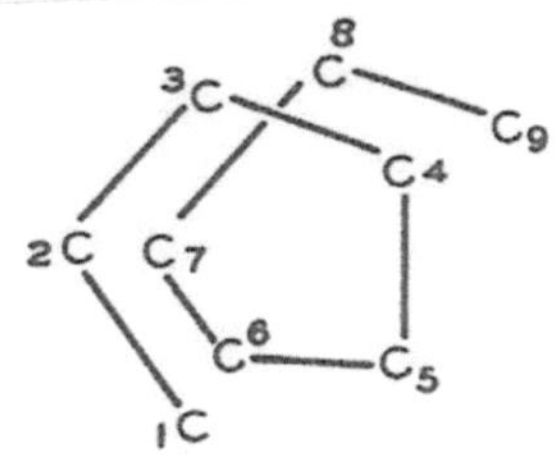

Figure 3. Idealized depiction of a long chain hydrocarbon coiled by virtue of London forces in a hydrophobic environment. Redrawn from [24]

The efficacy of the approach is considered later (*cf.* Table 8).

3. Removal of heavy metal ions with Octolig®

Remaining sections of this chapter are concerned with research involving the commercially available supported chelating agent called Octolig® because large quantities were available for a very reasonable cost (a wholesale price prior to 2008 was $40 per kilogram), and a collaboration was established with Robert Alldredge on the basis of mutual interest in the supported chelators, as noted elsewhere [10].

Examples of the removal of heavy metal ions by Octolig® may be found in three areas: patents, company literature, and refereed journals.

3.1. Patent literature

Lindoy and Eaglin [14] provided useful information about their experiments demonstrating the efficacy of removal of representative transition metal ions.

3.2. Company literature

Several examples are available in literature provided by Metre-General, Inc. [28].

A *first example* involved a Colorado plating shop with numerous plating lines, removal of Cr, Cu, Ni, and Zn was a matter of concern. The existing precipitation process did not consistently reduce heavy metal concentrations to less than required limits. By use of an Octolig® column chromatography system (cf. Figure 4), the plating shop was able to recycle the treated water and reduce the fresh water usage from 18,000 GPD (gallons per day) to 8,000 GPD [28].

Figure 4. Application example –The Octolig® ENVIRO-4000 of Metre-General, Inc.

Specifications included two 56-gallon HDPE tanks, a 120 V pump, on/off float switch, rotameter with flow-rate adjustment, two pre-filters, and four filters with refillable Octolig® cartridges. Size is 52"x52" with a 68 gallon spill containment. From [28] and reproduced with permission.

The results of the study are summarized in part in Table 2 and demonstrate two useful consequences. The first result was an effective removal of nuisance transition metal ions: 98% removal of zinc ion, 93% removal of nickel(II), 73% removal of copper(II) species. The second result, as a consequence, was the ability to reduce the fresh water intake by recycling reducing water usage by 67%.

Element	**Cr**	**Cu**	**Ni**	**Zn**
Untreated water	0.30	0.80	0.60	4.20
Treated water	N/D	0.21	0.04	0.0

Table 2. Effect of Octolig® treatment on concentrations, ppm, of selected metals [28]

A *second example* involved a similar unit installed at a gold mine located in mountains west of Denver, Colorado [27].The chief contaminant in the mine drainage water was zinc, and about 4 million gallons of mine drainage waster was discharged yearly at the gold mine. A two-fold treatment was tested. First, water flowing into a settling pond (33 ft. by 40 ft., 3 ft. deep) was crudely treated with lime. Second, the overflow was pumped through pre-filters then into an Octolig® MRP (metal removal plant) unit. In a pilot study, treated water contained only 0.02 ppm zinc. It was calculated that costs of the installation were less than $0.70 per 100 gallons of drainage water [28]. And probably one should remember the value of water in certain arid parts of the western United States.

A *third example* [28] involved a Superfund site. Octolig® was used in a bench test of material from the Liberty Industrial Finishing Superfund site, a 30-acre parcel in the village of Farmingdale, Nassau County, New York state. At this location, Republic Aviation produced military aircraft for the US government from the late 1930s until the 1960s. Operations involved electroplating aircraft parts. The area became contaminated with solvents that were used for cleaning parts as well as solutions of cadmium and chromium. Waste material was dumped into sand pits on the company property during the 1930s and 1940s. As a consequence, ground water became contaminated with low levels of solvents and low levels of cadmium and chromium, ca.0.300 mg/L (ppm). US Environmental Protection Agency designated the area as a Superfund site in the 1990s [28],

Metre-General, Inc. tested undiluted groundwater samples as received and at two dilutions to test the efficiency for removal at different influent concentrations. Dilute organics were removed by charcoal column chromatography. Heavy metals were removed by Octolig® column chromatography The removal system was one like that shown in Figure 4. Some 10 columns made of HDPE plastic about 35 in (89 cm) in diameter held 300 kg of Octolig®21. The columns were arranged in two banks of five each operated in parallel and designed to treat 300 GPM (gallons per minute). The results in Table 3 indicate that concentrations of metals of concern were reduced to below required limits.

The data also indicate that Octolig® can have a long "recycling life" because the heavy metal ions were removed, not comparatively innocuous ones like calcium that affect TDS (Total Dissolved Salts). The contrast between removal of cadmium or chromium versus non-removal of calcium ion shows the advantage of being able to design supported chelators for specific purposes. Supported iminodiacetic acid, for example would have removed cadmium, but also calcium ions as well, reducing in this instance the useful capacity of this material..

Dilution	Element	Influent Concentration, ppm	Effluent Concentration, ppm
None	Cd	0.280	0.029
1:5	Cd	0.061	0.007
1:10	Cd	0.032	<0.005
None	Cr	0.090	<0.005
1:5	Cr	0.023	<0.005
1:10	Cr	0.010	<0.005
None	Ca	25.0	25.0
1:5	Ca	5.1	4.8
1:10	Ca	2.7	2.5

From [28] and used with permission

Table 3. Summary of results of bench-scale treatment of Liberty Industrial Finishing Superfund Site.

3.3. Refereed literature

Quantitative removal of uranyl ion, UO_2^{++}, from aqueous solutions (well water)was demonstrated using the standard chromatography technique. Solutions of uranyl acetate (50 ppm)

were quantitatively removed by Ferrilig, Thorilig, or Octolig® [29]. Uranium is a contaminant of the mineral apatite in Florida that is a basis of the phosphate industry. Uranium can also contaminate sources of drinking water in certain areas of Colorado, and is a matter of concern for water supplies for small towns [29].

Gao and co-workers measured the absorption properties of an Octolig®- like material [13] by a batch and a flow methods. Quantitative reaction was reported, and the absorbing ability of the PEI-silica material followed the order of $Cu^{2+} > Cd^{2+} > Zn^{2+}$ at a pH of 6-7 [13]. They also measured the saturated absorption uptake and reported values for copper(II) of 25.95 mg/g and 50.01 mg/g, respectively, for static and dynamic conditions [13].

4. Removal of anions with Octolig® and metal derivatives of Octolig®

Some initial studies in our laboratory were concerned with removal of nuisance species by means of chromatography with Ferrilig, the iron(III) derivative of Octolig ® These species, existing as anions, were arsenate and arsenite, chromium as chromate, molybdenum as molybdenum(VI), and selenium as selenite and selenite species. All four are nuisance species in the western United States as well as elsewhere. Molybdenum is an essential element, whose compounds are useful, but it is a nuisance in areas of molybdenum mining when mining runoff water or processing water in ponds becomes a disposal problem. The initial focus was on removing arsenic species by chromatography, and a specific focus was on the iron(III) derivative of Octolig® (named "Ferrilig") because of the known insolubility of ferric arsenate [30-32].

The synthesis of Ferrilig (IV, Eqn 5), originally described [30] has been studied with view toward improving the amount of iron taken up. The synthesis is summarized (Eqn. 4-6) [31]. Octolig® (II) was treated with aqueous ferrous sulfate under a nitrogen atmosphere. The product (material III Eqn. 4, 5) after spontaneous oxidation and production of hydroxide ion (Eqn. 6) is termed Ferrilig (material IV). The structure of Octolig® was that given in the company literature [28].

$$\underset{\text{(II)}}{\text{-0}_3\text{Si-O -Si} - \text{CH}_2\text{CH}_2\text{CH}_2\text{CH}_2\text{NHCH}_2\text{CH}_2\left[\text{NH CH}_2\text{CH}_2\right]\text{n NH}_2} + \text{Fe}^{2+} \rightarrow \text{III} \qquad (4)$$

$$\text{III} + \text{O}_2 \rightarrow \underset{\text{(IV)}}{\text{-0}_3\text{Si-O-Si -CH}_2\text{CH}_2\text{CH}_2\text{CH}_2\text{NHCH}_2\text{CH}_2\left[\text{NH CH}_2\text{CH}_2\text{-}\right]\text{Fe(III)}} \qquad (5)$$

$$e^- + O_2 + H_2O \rightarrow 2\,OH^- \qquad (6)$$

Oxidation of the ferrous form (material III, green) to the ferric form (material IV, rust brown) occurs spontaneously in the presence of air, e.g., as the wet sample is standing exposed to the air.

http://dx.doi.org/10.5772/55521

The role of coordination in reducing the oxidation potential of iron(II) is well known, and was noted by Moeller [33], a process that is enhanced when the coordinating agent is a chelating agent. Thus the oxidation potential for hydrated ferrous-ferric species is - 0.771V; whereas the value in the presence of oxalate ion is -0.02 V [34]. Octolig® has a plethora of chelating species, i. e., ethylenediimino moieties or extended ethylenediamines, that should be capable of lowering the oxidation potential of coordinated iron(II). Accordingly, the ease of oxidation should hardly be surprising. Nevertheless, it was surely interesting to note and watch. Species II was white, species III was green, and species IV was rust-brown.

Considering the effectiveness of Ferrilig, the study of other metal derivatives ("metalloligs") was effected using what may be described as facile syntheses. The metals used were copper(II), cobalt(II), nickel(II), manganese (II), and thorium(IV) [32]. An exhaustive study can not be claimed, e.g., for all metalloligs and all anions. But all six metalloligs exhibited 99% removal of arsenic by means of column chromatography using 280 x a10^{-3} ppm As as Na_2HAs0_4 [32].Other anions were tested using various metalloligs, and quantitative removal (98-99%) was achieved for nitrate, nitrite, phosphate, sulfate, and fluoride ions in deionized water [31, 32, 35].

A *standard test* for removal ions by chromatography involved the following: A Spectra/chron peristaltic pump was used to deliver aqueous samples to a chromatography column, 2 cm (id) by 31cm and equipped with a glass frit and a Teflon stopcock. The column was packed with about 22 cm of Octolig® or other solid. Before packing, the solid was suspended in water, swirled, and the fines were decanted, a process that was repeated until no fines were observed. Water samples were chromatographed using a rate of 10 mL/min. Usually, the first three or four 50-mL aliquots of effluent were discarded, and later ones were used for analysis (Table 4). Total dissolved solids were measured, and used as a guide to assess a state of equilibrium [29-31, 35-38].

Element	Form	Sample	Initial Concentration, ppm	% Removal
As	Na_2HAsO_4	Well water	280x 10^{-3}	99.3
Cr	Na_2CrO_4	DI water	50.6	95.5
Mo	$(NH_4)_6\ Mo_7O_{24}{\bullet}4H_2O$	DI water	50.7	94.7
Se	Na_2SeO_3	Well water	258	99.9

Table 4. Effect of column chromatography of nuisance species using Ferrilig [31]

In a subsequent study, some attention was focused on the use of Cuprilig, obtained by a truly facile synthesis by shaking a suspension of Octolig® in deionized water and a standard solution of copper sulfate in deionized water. Cuprilig was tested for removal of perchlorate ion, which is a serious problem in certain areas, most notably in Rialto, California where one source of well water contained 10,000 ppb perchlorate. This remarkable concentration was probably a consequence of a plume of contaminated water, owing to proximity to a facility that produced ammonium perchlorate, the propellant for the sidewinder missile [37]. Obviously it was of interest to determine whether a metallolig, e.g., Cuprilig, could remove perchlorate ion from aqueous solutions using column chromatography, and it was a success [37]. Using perchlorate solution, the effluent was below detection limit, i.e., <1 ppb. But success also raised the question of a "control" experiment, one kind being to eliminate the role of the metal ion, and use only Octolig®. Doing so, the results were equally good, i.e., > 99% removal with Octolig® alone using deionized water or well water [37].

Clearly a different model for understanding the removal process was in order. And the current hypothesis is that nitrogens are protonated at low pH values probably as well, even somewhat above the pH at which Octolig® should not used. A working hypothesis consistent with Figure 5 is that at least two important considerations are involved: (1) The nitrogens must be protonated, and (2) species to be removed must be anions or if weak acids convertible to anions [39].

Further studies [35, 37, 38] have found that simple anions can be quantitatively removed by Octolig ® at reasonable values of pH, and the order of removal is consistent with an ionic model of anions being attracted to protonated moieties of the Octolig® as represented (Figure 5).

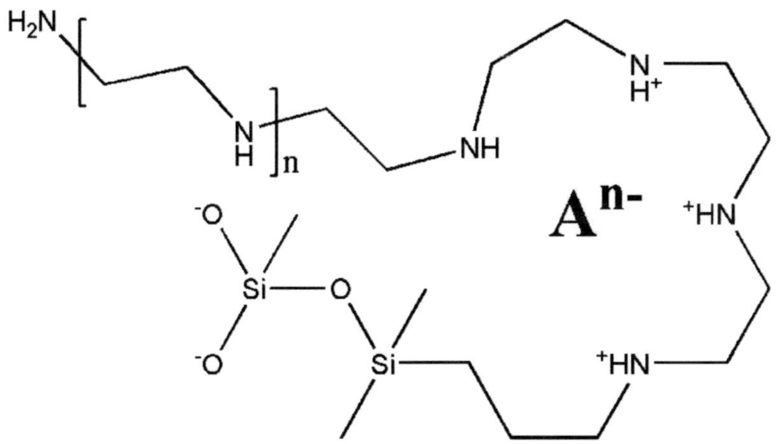

Figure 5. Proposed structure of Octolig® -anion (A-) interaction [38]. Reproduced with permissionof the publisher

A range of compounds can be quantitatively removed provided the pKa of the material is less than about 8 [40]. Above that value, the per cent removal was less than 20% [40]. This is demonstrated in Figure 6. It may be presumed that for those compounds having pKa values greater than 7+, the anion concentration vs. the degree of protonation of Octolig® reaches an unfavorable balance. One substance of special interest is BPA (*vide infra*).

Compounds in order of increasing pKa values (in parentheses): Amoxicillin (2.4), Eosin Y (2.7), Lissamine Green (~3), Erythrosine (3.6), Rose Bengal (3.9), 4-nitrophenol (7.15), 2-nitrophenol (7.22), 3-nitrophenol (8.36), 4-tert-butylphenol (10.16) 4-isoproplphenol (10.19). From [47] used with permission

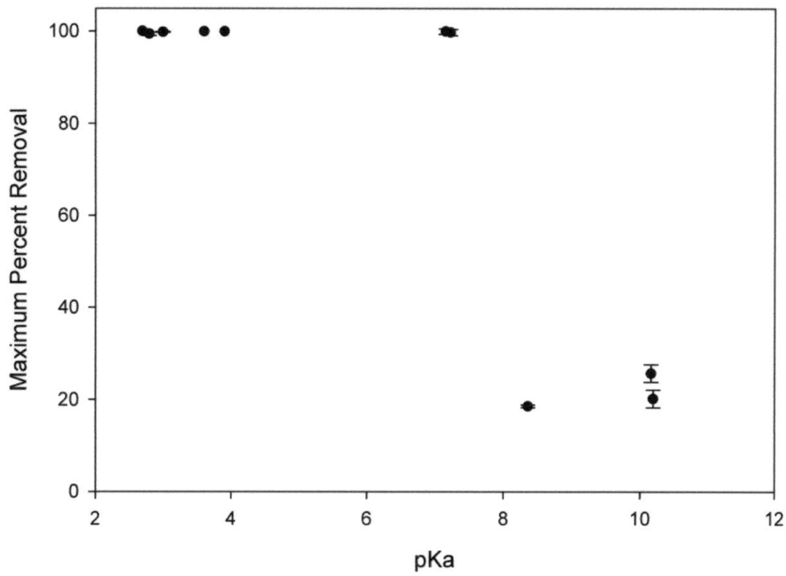

Figure 6. Maximum per cent removed as a function of the pKa for the compound under study [47]

5. Control of pharmaceuticals and other large anions

Three diverse groups of compounds can be usefully considered for their impact, but the need to be able to separate them from aqueous solution: Pharmaceuticals, BPA, and food dyes. These will be considered in succession.

5.1. Pharmaceuticals

The range of anions that can be quantitatively removed by Octolig® and column chromatography include some significant pharmaceuticals. These substances are chemical compounds (inorganic or organic) that can be used in the diagnosis, mitigation, treatment, or prevention of a disease [41,42]. Unfortunately, because of their usefulness and the magnitude of production they can represent a disposal problem [41-43]. Pharmaceuticals have an impressively wide range of applications – human medicine, veterinary medicine, aquaculture, livestock production, and agriculture. [43]. A recent review [42] quoted a statistic that of the 16,200 tons of antibiotics produced in the United States in 2000, about 70-% was used for livestock [40]. Unfortunately, about 75% of the antibiotics involved was not absorbed and was delivered to the environment in the form of urine and feces [41].

On the positive side, a study using Octolig® as a separating agent showed that Amoxicillin (Figure7), a very popular antibiotic in the United States was among those quantitatively removed [44], as noted in Table 5. This a result that is consistent with the information provided in Figure 6. It is also noteworthy that results for DI (deionized) water and well water from the Floridan Aquifer were similar. The lack of a matrix effect, at least for these two solvents may be promising if one considers that hospitals can represent a significant source of un-metabolized drugs as well as metabolic products of these drugs /in waste products. The concern raised more recently is the fate of such waste materials. One may consider an analogy, i.e., radiator shops in certain localities must be careful not to release zinc ion water laving the establishment lest the concentration affect bacterial used in sewage treatment. Will a similar concern arise with respect to hospitals? The possible causes for concern were raised in a review on the topic [43].

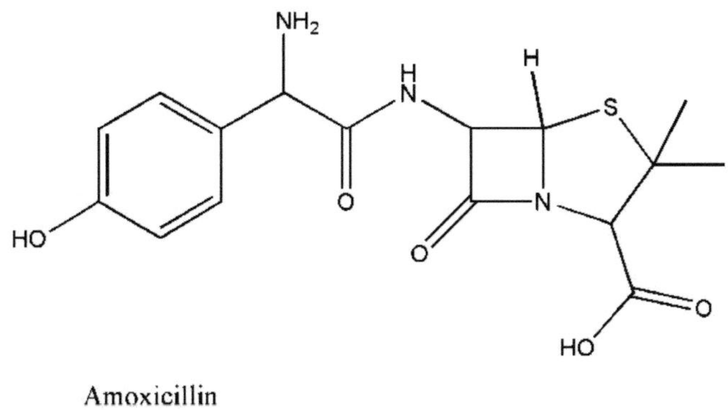

Figure 7. Structure of Amoxicillin, pK = 2.4

Matrix	TDS, ppm	pH	Stock Conc. (106 M)	% removed
DI water	4	6.17	639	99.4±0.4
	3	6.19	1230	99.8±0.3
	2	5.97	741	98.8±0.6
Well water	119	6.56	750	99.2±0.4
	153	7.12	912	96.9±1.2

Table 5. Passage of aqueous Amoxicillin samples over a 3.0 (id) chromatography column packed with ~130 mL of Octolig® at a flow rate of 10 mL/min (50-mL aliquots were collected and concentrations of fractions 4-10 were measured spectrophotometrically) [44]

5.2. PBA

Another large molecule of potential interest would be anions derived from Bisphenolacetone (Figure 8) synthesis (Eqn 7), notable reactions to produce polycarbonate (Eqn. 8) or epoxy resins (Eqn. 9). As reviewed recently elsewhere [45]. Bisphenol A (BPA), the synthesis of which is shown (Eqn. 7), is a component in the synthesis of polycarbonate plastics and epoxy resins (Eqn. 8, 9) that have a variety of significant uses.

http://dx.doi.org/10.5772/55521

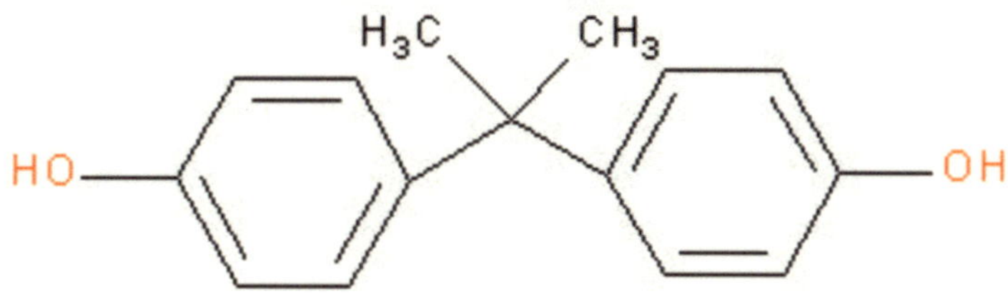

Figure 8. Bisphenol A, BPA. pK values = 9.59, 11.3

$$\underset{\text{phenol}}{2\,C_6H_5OH} + \underset{\text{acetone}}{(CH_3)_2C{=}O} \rightarrow \underset{\text{BPA}}{4\text{-}HOC_6H_4C(CH_3)_2C_6H_4OH\text{-}4} + H_2O \tag{7}$$

$$BPA + \underset{\text{phosgene}}{Cl_2C{=}O} \rightarrow \underset{\text{polycarbonate}}{\text{-}\left[O\text{-}C_6H_4C(CH_3)_2C_6H_4O(C{=}O)\text{-}\right]_n} \tag{8}$$

$$BPA + \underset{\text{epichlorohydrin}}{CH_2CHOCH_2Cl} \rightarrow \underset{\text{epoxy resin}}{\text{-}\left[O\text{-}C_6H_4C(CH_3)_2C_6H_4O\text{-}CH_2C(OH)CH_2\text{-}\right]_n} \tag{9}$$

As noted in the review [45], these plastics can fail with age, and BPA or other substances possessing estrogenic activity (EA) can be released. The review noted other points of significance (with documentation provided in the review) including:

- BPA is also present in our bodies in detectable amounts in our blood stream.
- It is present in the rivers and estuaries in detectable amounts, despite its low solubility in water.
- Though BPA may have a short half-life in soil, the ubiquity of the substance provides a continuous supply in the environment.
- A significant concern arises as to the toxicity of this material, which because of the ubiquity of the material and the uncertainty of the toxicity has become a matter of concern and significant debate, which seemingly leads to three choices: ban, restrict, or ignore [45, 46].
- Two better approaches are avoidance of exposure to EA-containing polycarbonates or treatment of EA-containing polycarbonates.

The potential control of EA materials in plastics needs to be considered because the volume of plastics used annually means that there will be no sudden cessation of use. But there is reason for optimism, based on a recent study. Results of a survey of 455 commercially available plastic products for release of estrogenic active (EA) material gives pause for the thoroughness as well as for the implications [46]. These workers discovered that EA materials could be removed by extraction with two solvents, e.g. ethanol and saline, and the potential for control of EA materials in plastics may be considered.

The hypothesis that anions of BPA might be removed by column chromatography with Octolig® was considered but not pursued because of the warnings in a MSDS (Material Safety Data Sheet) about hazards of BPA and because the pKa values of BPA were listed as 9.59 and 11.3.

The expectation of success was limited, based on the hypothesis and the prediction based on available data (Figure 6).

5.3. Dyes

As Rosales and co-workers noted [48], vast amounts of chemical dyes (around 10^6 tons) are made annually worldwide [49]. Dye-containing effluents can make their way into runoff and wastewater, eventually settling in the soil. As these workers noted [48], with textile industries, as much as 50% of the dyes can be lost and disposed in effluents [50]. These dyes can have adverse effects on the environment and ecosystems they pollute. Previous extraction methods have had limited success in removal from soil, and a recent approach involved the use of Fenton's reagent with electrochemistry[48] testing removal of Lissamine Green (Figure 9) from a pseudo-soil matrix (kaolin).

Martin and Nabar [51] noted that previous studies [39] had demonstrated the ability to remove Lissamine Green from aqueous solutions, using column chromatography with Octolig® so it might be cheaper to extract Lissamine Green from soil using hot water, then remove the Lissamine by column chromatography with Octolig®. They were successful with kaolin and montmorillonite, but discovered that mixtures of clay and peat were less successful depending on the amount of peat present, then success of extraction decreasing linearly with the concentration of peat present. accordingly, the two step procedure would save same on electricity, depending on the availability of hot water and the type of soil present.

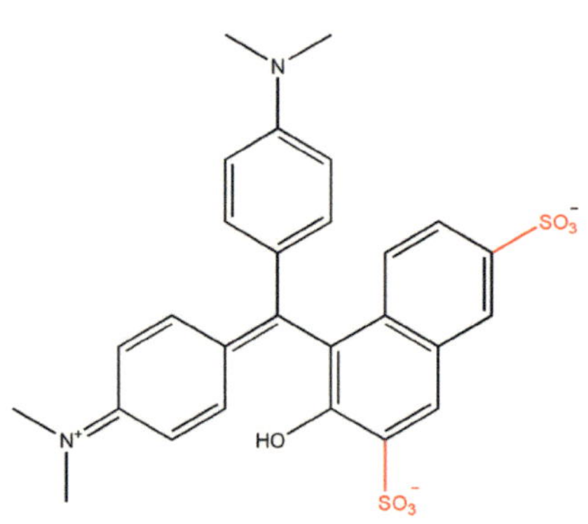

Lissamine Green B

Figure 9. Structure of Lissamine Green

5.4. Food dyes

Other dyes such as food dyes may be commonly present, but the amount actually used is uncertain, and some workers have been concerned with the impact of these commonplace substances.

Food dyes are among the commonplace aspects of our daily life that may need more scrutiny [52].. These artificial colors are added to food for several reasons, chiefly to make the food more appealing, or perhaps fool the consumer into thinking the food has fruit or other helpful ingredients [52]. As noted elsewhere [52], synthetic food dyes have no nutritional value, they have no health benefits, they are not preservatives, but they do make us feel good about eating the food.

One significant concern is the suspicion that for over three decades the dyes have not been safe for all consumers[52, 53] Specifically some dyes are suspected of being responsible for behavioral problems in children., including "short attention span, aggressiveness, impulsivity, distractibility[54]." Feingold is credited with proposing that food dyes induce or aggravate symptoms of hyperactivity in children [55], and suggesting a "Feingold diet" is one that eliminates artificial food colorings.

Conflicting results, however, made it difficult to determine whether a Feingold diet is beneficial, despite a number of studies that have been conducted that have led to the view that food dyes did impair performance of hyperactive children [54], or did not in a controlled study [56], or that it did and that a one-week experimental diet could be used to detect a "sub-group of children hyperactive from specific food dyes[57]."

One common food dye is FD&C No 1 (Figure 10), a dye which should be removable by column chromatography by Octolig® were this to be desired. The structure (Figure 10) showing sulfonate groups indicates on the basis of Figure 6 that a low pKa would be expected, as would be ease of separation.

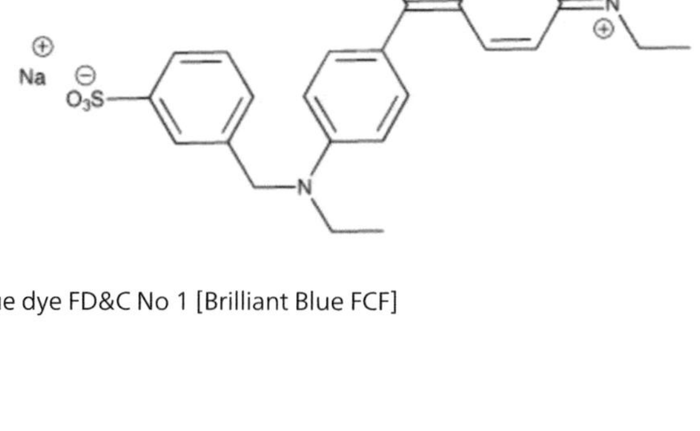

Figure 10. Structure of blue dye FD&C No 1 [Brilliant Blue FCF]

Using the *standard method*, noted earlier, quantitative removal (100.1+±0.04 %) of FD&C No 1 was achieved [58] This is one of seven food dyes approved for use in the United States under the Pure Food and Drug act of 1906 (abbreviated as FD&C), the other six dyes also appear to be easily removable by Octolig® based column chromatography based on an consideration of their structure.

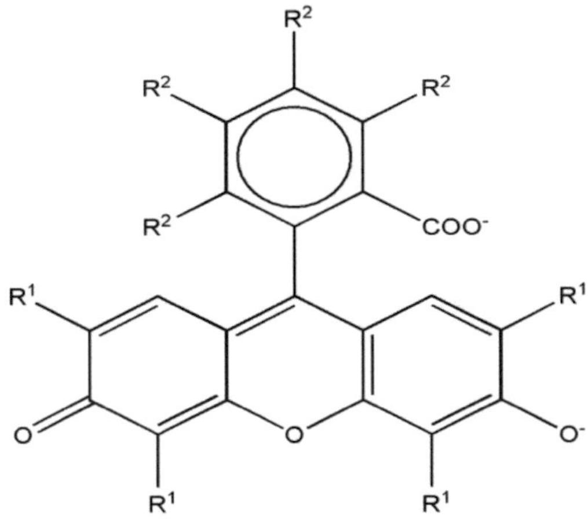

Figure 11. FiFD&C No 3., Erythrosine B, $R^1 = I$; $R^2 = H$

Matrix	pH	Concentration., µM	% Rermoved
DI water	8.66	78.8	99.9 ±0.0
Tap water	7.76	94.9	99.9 ± 0.0
Well water	8.33	105.5	98.6 ± 0.1

Table 6. Column chromatography of aqueous Erythrosine B samples over a 3.0 cm (id) column packed with ~130 mL of Octolig® at a flow rate of 10 mL/min (50-mL aliquots were collected and concentrations of fractions 4-10 were measured spectrophotometrically) [44]

Similarly, using our standard method for column chromatography, quantitative separation was obtained. for Erythrosine, but also for the other food dyes in contemporary use It is also notable that there was no significant matrix effect observed for DI, tap or well water.

6. Batch separations versus column chromatography using Octolig®

Removal of metals by immobilized ligands frequently involves a choice of two techniques, a batch method or column chromatography. The batch method can be faster when samples are taken from the supernatant for replicate analyses. This method also can establish when equilibrium (or stasis) has been established. One flaw is that the sample of solid may not be adequate to remove all the sample in the supernatant (which is an asset for measuring capacity of the solid). But there are at least three applications of the batch method that can be considered.`

One application was determining the time course for of the batch methods. The example (Figure 13) shows the time course of removal can be fairly rapid, and an estimate of the capacity is indicated by the "plateau phase."

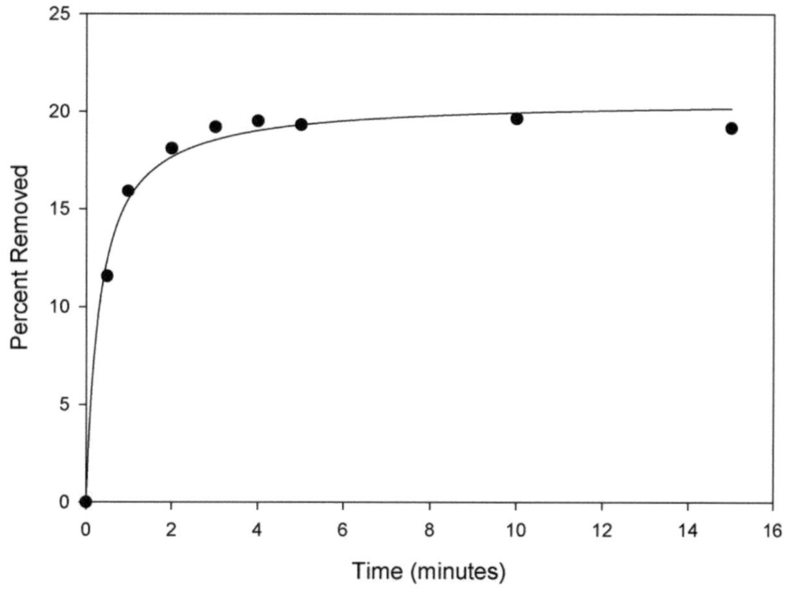

Figure 12. method with Octolig®: Percent removed by Octolig® as a function of time for aqueous 3-nitrophenol. Octolig® was suspended in 100 mL of aqueous 3-nitrophenol) and shaken (.at a rate of 240 rpm. Aliquot portions taken periodically as noted and analyzed. (Figure from [47], used with permission of the author).

A *second application* of the batch method was being able to evaluate a mechanism of sorption, as noted by Gao and co-workers [13].

A *third application* was that of comparison. Many in academe seem to have favored batch methods. In contrast, a valued colleague [9, 10] noted that information obtained from column chromatography was more applicable to the needs of industry. Column chromatography was used with Octolig® in practical applications as noted previously.

Accordingly, a series of experiments was designed and performed to evaluate comparisons of batch versus column chromatography (cf. Table 7). A *standard batch method* is presented here for the sake of comparison [59]. A sample of Octolig ® (5 g as received) was placed in a 250-mL Erlenmeyer flask covered with 100 mL of about 1600 ppm phosphate as NaH_2PO_4. The samples were placed in a gyrotory water bath and subjected to shaking (>170 rpm) overnight. At the end of the shaking period, an aliquot was removed, and diluted ca. 1:8 for phosphate. Mean and SD values were calculated. The result was subtracted from the initial phosphate concentration to determine the capacity, expressed as moles per kg of Octolig®.

Results of the study with Octolig®, summarized in Table 7 indicate that there was no statistically significant difference [59] between the two methods. Similar results were obtained for a capacity for arsenate [59]. The results (Table 7) were also used to see if there was a major difference between different preparations of Octolig® by comparing capacity for phosphate (Sample 1 vs. Sample 2),

Substrate	Anion	Chromatography*	Batch method*
Cuprilig	Phosphate	0.178±0.003	0.185±0.003
Octolig® (Sample 1)	Phosphate	0.304±0.02	0.299±0.004
Octolig® (Sample 2)	Phosphate	0.341±0.06	0.327±0.001

*Standard method (see text) was used

Table 7. Selected removal capacities (moles/kg) calculated for phosphate comparing the chromatography method versus a batch method (= 3) [59]

A *fourth application* is a convenient assessment of the removal of a transition metal by a supported chelator vs sorption on the substrate. It appears that the supported chelators were able to remove copper ion in a quantitative manner, but a goodly proportion (83%) was removed by sorption on silica gel. In contrast, a Linde molecular sieve (alone) with fairly defined pores removed about half of the copper ion through sorption.

Chelator	Support	% Removed
LIX® 54	silica gel	97.7±3.3
Oleoylacetone	silica gel	100±1.1
-----	silica gel	83.2±2.2
LIX® 54	Linde 3A	89.4±2.2
------	Linde 3A	42.4±5.5

Table 8. Extraction of 5 ppm copper from ammoniacal solution using supported chelating agents and supported chelators prepared by the corkscrew method as noted above. Modified from [26] Summary

Supported chelating agents can effectively satisfy three needs: concentration for analysis, removal from solution, and removal coupled with regeneration. Their use on a commercial scale has been demonstrated for many years with Chelex as well as Octolig® for removal of transition metals ions and other uses. It seems likely that Octolig® could be competitive with a highly selective ion-exchange resin and, perhaps, commercially competitive, but this remains to be demonstrated. The examples demonstrated indicate a number of chelating agents can be attached using one of three different methods, though covalent attachment seems the easiest and most assured. The range of separations --- metals ions, simple anions, pharmaceuticals, industrial dyes, and food dyes – seem impressive, and indicates the utility of separation by column chromatography using appropriately supported chelating agents.

Acknowledgements

I am grateful to the collaborators/co-authors who worked with me over the past two decades. I have valued the association with Mr. Robert L. Alldredge (dec) and Mr. Mark H. Alldredge. I thank Mrs. Barbara B. Martin for helpful comments and encouragement. I am grateful for the encouragement of Ms. Viktorija Zgela.

Author details

Dean F. Martin

Address all correspondence to: dfmartin@usf.edu

Institute for Environmental Studies, Department of Chemistry-CHE, University of South Florida, Tampa, Florida, USA

References

[1] Chelex 100 and Chelex 220 Chelating Ion exchange Resin Instruction Manual, http://www.biorad.com/webmaster/pdfs/9184_Chelex.PDF) (accessed 26 May 2012).

[2] Ceo RN, Kazerouni MR, Rengan K. 1993. Chelex 100 as a Medium for Simple Extraction of DNA for PCR-based Typing from Forensic Material. Biotechnique 1993; 10 506-513.

[3] Pai SC, Whung PY, Lai RL. Preconcentration Efficiency for Chelex-100 Resin for Heavy Metals in Sea Water, parts 1 and 2 Analytica Chimica Acta 1998; 211 257-280.

[4] Florence TM, Batley GE. Trace Metal Species in Sea Water-I. Removal of Trace Meals from Seawater by a Chelating resin. Tantala 1976; 23 179.

[5] Figura P, McDuffie B. Characterization of the Calcium Form of Chelex-100 for Trace Metal Studies. Analytical Chemistry 1977; 49 1950-1953.

[6] Ryan DH, Weber JH. Comparison of Chelating Agents Immobilized on Glass with Chelex-100 for Removal and Preconcentration of Trace Copper. Tantala 1985; 32 859-863.

[7] Hill JM. Silica Gel as an Insoluble Carrier for the Preparation of Selective Chromatographic Absorbents. The Preparation of 8-Hydroxyquinoline Substituted Silica Gel for the Chelation Chromatography of Some Trace Metals. Journal of Chromatography 1973; 76 455-458.

[8] Irving HMNH, Williams RJP. The Stability of Transition-metal Complexes. Journal of the Chemical Society 1953;3192–3210. DOI:10.1039/JR9530003192.

[9] Alldredge RL 2004 Personal communication; cf [10].

[10] Martin DF, Martin BB. Robert L Alldredge, Chemical Engineer, Inventor, Entrepreneur. Technology and Innovation 2011; 13 203-21.

[11] Anon. Berkeley Pit http://en.wikipedia.org/wiki/Berkeley_Pit (accessed 18 June 2012).

[12] Anon. Berkley Pit Innovative Technologies Project, Phase II Metre General, Inc. demonstration. Mine Waste Technology Program Activity IV, Project7; MSE Technology Applications, Inc. and Montana Tech of the University of Montana, Butte, MT. prepared for USEPA IAD ID No. DW89938513-01-0. No date given..

[13] Gao B, An F, Liu K. Studies on Chelating Adsorption Properties of Novel Composite Material Polyethyleneimine/silica Gel for Heavy-metal Ions. Applied Surface Science 2006; 253 1946-1952.

[14] Lindoy L F, Eaglen P. Ion Complexation by Silica-immobilized Polyethyleneimens, U.S. Patent 5,190,660, Awarded March 2, 1993.

[15] Soliman EM. Synthesis and Metal Collecting Properties of Amines, Mono, Di, Tri, and Tetraamine Based on a Silica Gel Matrix. Analytical Letters 1997; 30 1739-1751.

[16] El-Ashgar, N M Column Extraction and Some Metal Ions by Diethylenetramine Polysiloxane Immobilized Ligand System. E-Journal of Chemistry 2008; 5 107-113.

[17] Martin DF, Bowe CA. Extraction of Metals by Mercaptans Attached to Silica Gel by Azeotropic Distillation. U.S. Patent 7,193,114. Awarded March 20, 2007.

[18] Bowe CA, Krikorian N, Martin DF. Extraction of Heavy Metals Using Modified Montmorillonite KSF. Florida Scientist 2004; 67 74-79.

[19] Krikorian N, Martin, DF. Extraction of Selected Heavy Metals Using Modified Clays. Journal of Environmental Science and Health, Part A 2005; A40 601-607.

[20] Bowe, CA.; Martin, DF. Extraction of Heavy Metals by 2-Mercaptoethoxy Groups Attached to Silica Gel. Journal of Environmental Science and Health, Part A 2004; A39 1479-1485.

[21] http://en./Wikipedia.org/wiki/Dean-Stark_apparatus (accessed 8 August 2012).

[22] Norris CW, Benson RF, Martin DF. Extraction of Heavy Metals by Resin-supported Chelators. Florida Scientist 1996; 59 174-18.

[23] Lee KS, Lee W, Lee DW. Selective Separation of Metal Ions by a Chelating-loaded Anion Exchanger. Analytical Chemistry 1978; 50 255-258.

[24] Benson RF, Martin DF. Supported Chelating Agents. The Corkscrew Model. Journal of Environmental Science and Health, Part A 1992; 27 2191-2197.

[25] Pooré DD, Benson RF, Martin DF. Removal of Heavy Metal Ions From Aqueous Solutions Using Dithiooximides Supported on Silica Gel. Journal of Environmental Science and Health, Part A 1996; A31 2167-2172.

[26] Norris CD, Benson RF, Martin DF. Extraction of Heavy Metals by Chelators Attached to Silica Gel by a Corkscrew Mechanism. Journal of Environmental Science and Health, Part A 1997; A32: 123-132 (1997).

[27] Bowe CA, Benson RF, Martin DF. Extraction of Heavy Metals by Mercaptans Attached to Silica Gel by a Corkscrew Mechanism. Journal of Environmental Science and Health, Part A 2002; A37 1391-1398.

[28] Anon. Tomorrow's Solution Today. Metre-General, Inc. Frederick Colorado 2005 http://www.octolig.com (Accessed 19 July 2012).

[29] Martin, DF Alldredge R. Removal of Uranium(VI) from Water Samples Using Octolig® and Ferrilig, an Iron Derivative of Octolig®, a Supported Chelating Agent. Florida Scientist 2008 ; 71: 208-214.

[30] Martin DF, O'Donnell L, Martin BB, Alldredge R.. Removal of Aqueous Arsenite Using Iron-attached to Immobilized Ligands (IMLIGs). Journal of Environmental Science and Health 2007 42A, 97-102.

[31] Martin DF, O'Donnell L, Martin BB, Alldredge R L. Removal of Aqueous Nuisance Anions with Ferrilig. Journal of Environmental Science and Health 2008; 43A: 700-704.

[32] Martin DF, Aguinaldo JS, Kondis NP, Stull FW, O'Donnell LF, Martin BB, Alldredge R L. Comparison of Effectiveness of Removal of Nuisance Anions by Metalloligs, Metal Derivatives of Octolig®, Journal of Environmental Science and Health 2008; 43A 1296-1302.

[33] Moeller, T. Inorganic Chemistry. New York: John Wiley & Sons, Inc.; 1952.

[34] Lattimer, WM. The Oxidation States of the Elements and Their Potentials in Aqueous Solutions, 2nd ed., Englewood Cliffs, NJ, : Prentice Hall; 1952 223-227.

[35] Stull FW, Martin DF. Comparative Ease of Separation of Mixtures of Selected Nuisance Anions (Nitrate, Nitrite, Sulfate, Phosphate) using Octolig®. Journal of Environmental Science and Health 2009; 44 1551-1556.

[36] Martin DF, Martin BB, Alldredge, RL.. Arsenic, Nitrate, and Perchlorate in Water --- Dangers, Distribution, and Removal. Bulletin for the History of Chemistry 2008; 33 17-24.

[37] Martin DF, Kondis NP, Alldredge RL. Effectiveness of Removal of Aqueous Perchlorate by Cuprilig, a Copper(II) Derivative of Octolig. Journal of Environmental Science and Health Part A 2009; 44 188-191.

[38] Martin DF, Lizardi C L, Schulman E, Vo B, D. Wynn D., Removal of Selected Nuisance Anions by Octolig®, Journal of Environmental Science and Health, Part A 2010; 45 1145-1149.

[39] Chang, W-S, Martin DF, Small M, Use of model compounds to study removal of pharmaceuticals using Octolig® Technology and Innovation. 2010; 12 71-77.

[40] Alessio RJ, Li X, Martin DF. Removal of BPA model compounds and related substances by means of column chromatography using Octolig®. Journal of Environmental Science and Health 2012; 47 000-000.

[41] Kümmerer K. Significance of Antibiotics in the Environment. Journal of Antimicrobial. Chemotherapy 2003;52 5-7.

[42] Kümmerer, K., editor, Pharmaceuticals in the Environment. Sources, Fate, Effects and Risk. 3rd ed. Berlin/Heidelberg: Springer; 2008.

[43] Martin DF, Ward DR, Martin BB., Agricultural Pharmaceuticals In The Environment. A Need For Inventiveness, Technology and Innovation. 2010; 12 (3), 129-141.

[44] Chang, W-S, Martin DF, Small M. Removal of Selected Pharmaceuticals Using Octolig®, Technology and Innovation 2010; 12 143-152.

[45] Martin DF, Martin BB. Polycarbonates from Bisphenol A: A Good Invention Gone Awry? Technology and Innovation 2012; 14 7-15.

[46] Yang CZ, Yaniger SI, Jordan VC, Klein D J, Bittner GD. Most Plastic Products Release Estrogenic Chemicals: A Potential Health Problem That Can Be Solved. Environmental Health Perspectives 2011; 119 989-996.

[47] Alessio, RJ. Removal of Bisphenol A model compounds using Octolig®MS Thesis. University of South Florida; 2012.

[48] Rosales E, Pazos M, Longo MA.; Sanromán, MA. Influence of Operational Parameters on Electro-Fenton Degradation of Organic Pollutants from Soil. Journal of Environmental Science and Health. Part A 2009; 44 1104-1111.

[49] Sanroman M, Pazos M, Ricart M, Cameselle C Decolourisation of Textile Indigo Dye by DC Electric Current. Engineering Geology 2005,; 77 (3-4),253-61

[50] Zollinger H. Color of Organic Compounds. Color Chemistry: Synthesis, Properties, an Applications of Organic Dyes and Pigments, 3rd Ed.; Wiley-VCH: Zurich, 2003; 1, 3-146.

[51] Martin DF, Nabar N. Studies on the Removal of Lissamine Green from Soil and Comparison with Contemporary Approaches Journal of Environmental Science and Health. Part A 2012; 47 260-266

[52] Erickson B. Food dye debate resurfaces. Chemical and Engineering News 2011, April 18, pp 27-31.

[53] McCann D, Barrett A, Cooper A, Cumpler D, Dalen L, Grimshaw K, Kitchin E., Lok K, Porteous L, Prince E, Sonuga-Barke E, Warner JO, Stevenson J. Food Additives and Hyperactive Behavior in 3-year-old and 8/9-year-old Children in the Community: a Randomized, Double-blinded, Placebo-controlled Trial. Lancet. Published online September 6, 2007 DOI: 10.1016/S0140-6736(07)61306-3.

[54] Swanson JM, Kinsbourne, M. Food Dyes Impair Performance of Hyperactive Children on a Laboratory learning test. Science 1980; 207 1485-1487.

[55] Feingold BF. Why Your Child is Hyperactive, New York: Random House; 1975.

[56] Mattes JA, Gittleman R. Effects of Artificial Food Colorings in Children with Hyperactive Symptoms. Archives of General Psychiatry 1981; 39 714-718.

[57] Rappp DJ. Does Diet Affect Hyperactivity? Journal of Learning Disabilities 1978; 11 56-62

[58] Martin DF Alessio RJ, McCane CH. Removal of Synthetic Food Dyes in Aqueous Solutions by Octolig®, Journal of Environmental Science and Health. Part A. accepted

[59] Martin DF, Franz D. Comparison of Anion Percent Removal Capacities of Octolig® and Cuprilig, Journal of Environmental. Science and Health, Part A 2011; 46 1619-1624.

Chapter 6

A General Description of Apocynaceae Iridoids Chromatography

Ana Cláudia F. Amaral, Aline de S. Ramos,
José Luiz P. Ferreira, Arith R. dos Santos,
Deborah Q. Falcão, Bianca O. da Silva,
Debora T. Ohana and Jefferson Rocha de A. Silva

Additional information is available at the end of the chapter

http://dx.doi.org/10.5772/55784

1. Introduction

1.1. Iridoids

Iridoids are considered atypical monoterpenoid compounds, based on a methylcyclopentan-[C]-pyran skeleton, often fused to a six-membered oxygen ring consisting of ten, nine or in rare cases, eight carbon atoms (Figure 1a) [1, 2]. More than 2500 iridoid compounds have been described in nature, with structural differences related mainly to the degree and type of substitution in the cyclopentane ring skeleton [3]. Iridoids can be found in nature as secoiridoids (Figure 1b), a large group characterized by cleavage of the 7,8-bond on the cyclopentane ring, glycosides, mainly 1-*O*-glucosides, and nor-iridoids, originating from oxidative decarboxylation on C_{10} or C_{11} (Figure 1) [3, 4].

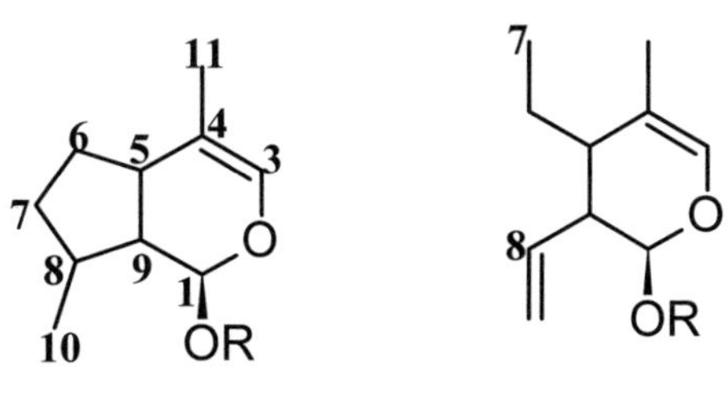

Figure 1. Basic skeleton a) iridoid; b) seco-iridoid (R=H or glucose)

Iridoids are derived from isoprene units, which are considered the universal building blocks of all terpenoids, formed through intermediates of the mevalonic acid (MVA) pathway in the citosol, and the novel 2-methyl-D-erythritol 4-phosphate (MEP) pathway in the plastids of plant cells [2, 5, 6]. The participation of two pathways in iridoid biosynthesis has not yet been clarified, but recent analyses have described the major role of the MEP in the yield of the source for the iridoid isoprene units when compared with the MVA pathway [7, 8, 9]. Iridoid biosynthesis shows two pathways, called route I and II, in which secoiridoids and carboxylated or decarboxylated iridoids are formed, respectively. Route I, considered the main pathway, is responsible for yielding the precursor of the carboxylic iridoids, from iridodial which is oxidized a iridotrial and subsequently converted to a series the iridoids, as occurs in loganin, secologanin, derived secoiridoids, and complex indole alkaloids. In route II, the presence of 8-epi-iridodial, 8-epi-iridotrial and 8-epi-deoxyloganic acid have been reported, forming a source of decarboxylated carbocyclic iridoids such as aucubin and catalpol [10, 3, 11, 12].

Iridoids have shown a broad range of biological activities, such as an antibacterial, antifungal, anti-inflammatory, antitumoral, hepatoprotective, cardioprotective, antioxidative, anti-protozoal and anti-insect properties [13, 14, 15, 16, 17, 18, 19]. *In vitro* activities inhibiting the hepatitis C virus, the differentiation of the adipocyte, and PPARα activation activities have been also described [20, 21].

The distribution of iridoids in the Eudicotyledoneae has potential usefulness in the taxonomy of the families, related to their presence in a restricted number of families. Iridoid are considered good chemotaxonomic markers of different taxononomic groups, and can be used, in combination with order, tribe and family, to establish the phylogenetic relationship [22, 23, 10, 3, 24, 25].

According to an update of the Angiosperm Phylogeny Group (APGIII) [28,29], the presence of iridoids has been reported in approximately fifty plant families, and can be considered as one of the synapomorphies of the Asterid clade. It is divided into Lamiids, which presents iridoids of the Gentianales, Garryales and Lamiales orders, and Campanulids, which presents secoiridoids of the Asterales and Dipsacales orders. The Gentianales order comprises five families: Apocynaceae, Gelsemiaceae, Gentianaceae, Loganiaceae and Rubiaceae (APG III, 2009), highlighted for the diversity of their iridoids [11, 29, 30].

1.2. Apocynaceae family

Apocynaceae is the most important family within this order, with ca 5000 species distributed worldwide. Seventy percent of the genus, and half of the species distributed in the Neotropical region, are found in the native Brazilian flora [31, 33]. Today, five subfamilies are described for the Apocynaceae family: Rauvolfioideae, Apocynoideae, Secamonoideae and Asclepia-doideae. The latter is the major subfamily of the Apocynaceae, and comprises approximately 3000 species divided into 172 genera, distributed mainly in Neotropic areas of South America, such as Brazil, where the highest diversity of the species has been found [32, 33, 34].

The iridoids class has significant distribution within Apocynaceae family, but is concentrated in just a few genera. The most representative of these are: *Plumeria, Himatanthus, Allamanda* and *Cerbera.* According to the traditional classification, the family comprises approximately 87 iridoids, the main ones being plumieride, plumericin and isoplumericin. A review of literature comprising works published on the identification of this class of constituents within Apocynaceae showed, as major natural sources, the following species (numbers in brackets indicate species as shown in Table 1) [36]:

(1). *Plumeria rubra* L. [*P. acuminata* W. T. Aiton; *P. acutifolia* Poir.; *P. bicolor* Ruiz & Pav.]; (2). *P. lancifolia* Müll. Arg.; (3). *P. acutifolia* Poir. [*P. rubra* L.]; (4). *P. alba* L. [*P. alba* var. *jacquiniana* A. DC.; *P. hypoleuca* Gasp.; *P. hypoleuca* var. *angustifólia* Gasp.]; (5). *P. bracteata* A. DC.; (6). *P. obtusifolia* Steud.; (7). *P. obtusa* L. [*P. multiflora* Standl.]; (8). *P. rubra* var. *alba;* (9). *P. multiflora* Standl.; (10). *Allamanda neriifolia* Hook. [*A. cathartica* var. *Schottii* (Pohl) L.H. Bailey & Raffill]; (11). *Allamanda cathartica* L. [*A. grandiflora* (Aubl.) Lam.; *A. schottii* Hook.]; (12). *Alstonia boonei* De Wild. [*A. congensis* Engl.]; (13). *Cerbera. manghas* L.; (14). *Alyxia reinwardtii* Blume; (15). *Alstonia scholaris* (L.) R.Br.; (16). *Himatanthus sucuuba* (Spruce ex Müll. Arg.) Woodson [*H. tarapotensis* (K. Schum. Ex Markgr.) Plumel; *Plumeria floribunda* Müll. Arg.; *P. tarapotensis* K. Schum. ex Markgr.]; (17). *Vinca* sp. L.; (18). *Thevetia peruviana* (Pers.) K. Schum. [*T. neriifolia* Juss. ex Steud.]; (19). *Himatanthus. phagedaenicus* (Mart.) Woodson [*Plumeria floribunda* var. *crassipes* Müll. Arg.; *P. lancifolia* var. *major* Müll. Arg.; *P. phagedaenica* Mart.]; (20). *Plumeria bicolor* Seem.; (21). *Plumeria acuminata* W.T. Aiton [*P. rubra* L.]; (22). *Cerbera odollam* Gaertn.; (23). A*llamanda. schottii* Pohl [*A. brasiliensis* Schott ex Pohl; *A. cathartica* Schrad; *A. neriifolia* Hook.]; (24). *Himatanthus articulatus* (Vahl) Woodson [*H. rigidus* Wild. Ex Roem. & Schult.; *Plumeria articulata* Vahl; *P. drastica* Mart.; *P. microcalyx* Standl.]; (25). *Rauwolfia grandiflora* Mart. Ex A. DC.; (26). *Plumeria inodora* Jacq. [*P. alba* L.; *P. alba* var. *fragrans* Kunth; *P. alba* var. *fragrantissima* G. Don; *P. alba* var. *inodora* (Jacq.) G. Don]; (27). *Himatanthus bracteatus* (A. DC.) Woodson; (28). *Himatanthus stenophyllus* Plumel; (29). *Himatanthus fallax* (Müll. Arg.) Plumel [*Plumeria fallax* Müll. Arg.]; (30). *Himatanthus obovatus* (Müll. Arg.) Woodson; (31). *Allamanda doniana* Müll. Arg.; (32). *Catharanthus roseus* (L.) G. Don; (33). *Nerium indicum* Mill. [*N. oleander* L.]; (34). *Alstonia macrophylla* Wall. ex G. Don; (35). *Winchia calophylla* A. DC.

1.3. Iridoids of Apocynaceae family

IRIDOIDS	SPECIES	PLANT MATERIAL	REF.
6''*O*-acetylplumieride *p-E*-coumarate	7: *Plumeria obtusa*	7:leaves	7: [37]; 7: [38]
6''*O*-acetylplumieride *p-Z*-coumarate	7: *Plumeria obtusa*	7:leaves	7: [37]; 7: [38]
13-*O*-acetylplumieride	10: *Allamanda neriifolia*	10:stem; 10:leaves	10: [39]; 10: [40]
allamancin	10: *Allamanda neriifolia*	10:stem	10: [39]; 10: [40]
allamandicin	10: *Allamanda neriifolia*	10:stem	10: [39]; 10: [41]; 10: [42]
allamandin	1: *Plumeria rubra;* 10: *A. neriifolia;* 11: *A. cathartica;*	1:stem bark; 19,23:stem; 10,11:leaves; 11,16:root	1: [43] ; 1: [44]; 1: [48]; 1: [49]; 10: [39]; 11: [41];

IRIDOIDS	SPECIES	PLANT MATERIAL	REF.
	23: *A. schottii;*16: *Himatanthus sucuuba;*19: *H. phagedaenicus;*		11: [42]; 16: [50]; 16: [51]; 19: [45]; 19: [46]; 23: [47]
allamanoid	10: *Allamanda neriifolia*	10:aerial parts	10: [52]; 10: [53]
allamcidin A	1: *Plumeria rubra;*10: *Allamanda neriifolia*	1:bark; 10:leaves	1: [43]; 1: [44]; 10: [39]; 10: [40]
allamcidin B	1: *Plumeria rubra;* 10: *Allamanda neriifolia*	1:bark; 10:leaves	1: [43]; 1: [44]; 10: [39]; 10: [40]
allamcidin B β-D-glucose	10: *Allamanda neriifolia*	10:stem	10: [39]; 10: [40]
allamcin	1: *Plumeria rubra;*10: Allamanda neriifolia; 23: A. schottii	1:stem bark; 10:leaves; 23: stem	1: [43]; 1: [44]; 1: [48]; 10: [39]; 10: [40]; 23: [47]
allamdin	11: *Allamanda cathartica*	11:root	11: [41]; 11: [42]
allaneroside	10: *Allamanda neriifolia*	10:leaves	10: [40]; 10: [54]
3,10-bis-*O*-allosylcerberidol	13: *Cerbera manghas;* 22: *C. odollam*	13,22:leaves	13: [40]; 13: [56]; 13,22: [55]
3-*O*-allosylcerberidol	13: *Cerbera manghas*; 22: *C. odollam*	13,22:leaves	13: [40]; 13: [56]; 13,22: [55]
3-*O*-allosylcyclocerberidol	13: *Cerbera manghas*; 22: *C. odollam*	13,22:leaves	13: [40]; 13: [56]; 13,22: [55]
3-*O*-allosylepoxycerberidol	13: *Cerbera manghas*; 22: *C. odollam*	13,22:leaves	13: [40]; 13: [56]; 13,22: [55]
alstonoside	15: *Alstonia scholaris*	15:stem	15: [57]; 15: [58]
alyxialactone	14: *Alyxia reinwardtii;* 15: *Alstonia scholaris*	14:leaves; 15:bark	14: [59]; 15: [60]
10-*O*-benzoyltheveside	13: *Cerbera manghas*	13:leaves	13: [56]; 13: [61]
10-*O*-benzoyltheviridoside	13: *Cerbera manghas;* 18: *Thevetia peruviana*	13,18:leaves	13: [56]; 13,18: [62]
boonein	12: *Alstonia boonei; 25: Rauwolfia grandiflora*	12: root; 25:bark	12: [40]; 12: [63]; 12: [64]; 25: [1]
13-*O*-caffeoylplumieride	1: *Plumeria rubra;* 3: *P. acutifolia*	1,3:root	1: [65]; 3: [38]; 3: [40]; 3: [43]; 3: [66]
10-carboxyloganin	13: *Cerbera manghas*	13: leaves	13: [56]; 13: [62]
cerberic acid	13: *Cerbera manghas*	13:bark	13: [40]; 13: [56]; 13: [67]
cerberidol	13: *Cerbera manghas;* 22: *C. odollam*	13,22:leaves	13: [40]; 13: [56]; 13,22 [55]
cerberinic acid	13: *Cerbera manghas*	13:bark	13: [40]; 13: [56]; 13: [67]

IRIDOIDS	SPECIES	PLANT MATERIAL	REF.
cerbinal	13: *Cerbera manghas*	13:bark; 13:cortex	13: [56]; 13: [67]; 13: [68]
champalinin ([(*E*)-*p*-methoxycinnamoyloxy] plumieride)	7: *Plumeria obtusa*	7:leaves; 7:stem bark	7: [43]; 7: [69]
cyclocerberidol	13: *Cerbera manghas;* 22: *C. odollam*	13,22:leaves	13: [40]; 13: [56]; 13,22: [55]
cyclocerberidol-3-*O*-*β*-D-glucoside	13: *Cerbera manghas*	13:leaves	13: [56]; 13: [62]
deglucosylplumieride	10: *Allamanda neriifolia*	10:stem	10: [39]
dehydrogardenoside-10	10: *Allamanda neriifolia*	10: leaves	10: [39]
10-dehydrogeniposide	13: *Cerbera manghas*	13:leaves	13: [56]; 13: [61]
15-demethylisoplumieride	1: *Plumeria rubra;*16: *Himatanthus sucuuba*	1,16:bark; 16:latex	1: [43]; 1,16: [53]; 1,16: [70]; 16: [51]
15-demethylplumieride	1: *Plumeria rubra*; 3: *P. acutifolia*; 4: *P. alba*; 7: *P. obtusa*; 16: *Himatanthus sucuuba*	1,16:bark; 3,4:leaves; 4:root; 7:aerial parts; 16:latex	1: [44]; 1,4: [43]; 3: [71]; 7: [72]; 16: [51]; 16: [73]
13-deoxyplumieride	1: *Plumeria rubra;* 3: *P. acutifolia*	1,3:root	1: [65]; 3: [38]; 3: [40]; 3: [43]; 3: [66]
β-dihydroplumericinic acid	1: *Plumeria rubra*	1:root	1: [38]; 1: [42]; 1: [43]; 1: [74]
β-dihydroplumericin	1: *Plumeria rubra;* 16: *Himatanthus sucuuba*	1,16:root; 1:stem bark	1: [38]; 1: [42]; 1: [43]; 1: [48]; 1: [74]; 16: [51]
epiplumeridoid C	1: *Plumeria rubra*	1: stem bark	1: [48]
epoxycerberidol	13: *Cerbera manghas;* 22: *C. odollam*	13,22:leaves	13: [40]; 13: [56]; 13,22: [55]
epoxycerberidol-3-*O*-*β*-D-glucoside	13: *Cerbera manghas*	13:leaves	13: [56]; 13: [62]
fulvoplumierin	1: *Plumeria rubra* L., 3: *P. acutufolia, 5: P. bracteata*, 8: *P. rubra* var. *alba;* 21: *P. acuminata;* 16: *Himatanthus sucuuba*	1,3,8,21:stem bark; 1,16:bark ; 1:root, 1:descorticated stem, 3:rind, 5:root cortex	1: [43]; 1: [48]; 1: [65]; 1: [74]; 1: [75]; 1: [76]; 1: [77]; 1,3,5,8: [38]; 1,3,21: [44]; 3: [78]; 3: [79]; 3: [80]; 3,8: [42]; 3,8: [83]; 5: [81]; 8: [82]; 16: [51]
gardenoside	10: *Allamanda neriifolia*	10:stem;10:leaves	10: [39]
isoallamandicin	10: *Allamanda neriifolia*	10:stem	10: [39]; 10: [40]

IRIDOIDS	SPECIES	PLANT MATERIAL	REF.
isoboonein	15: *Alstonia scholaris;* 25: *Rauwolfia grandiflora*	15,25:bark	15: [60]; 25: [1]
isoplumericin	1: *Plumeria rubra*, 3: *P. acutifolia;* 4: *P. alba*, 7: *P. obtusa;*20: *P. bicolor;* 10: *Allamanda neriifolia;*11: *A. cathartica;* 23: *A. schottii;* 31: *A. doniana;* 16: *Himatanthus sucuuba;*19: *H. phagedaenicus;* 24: *H. articulatus;* 29: *H. fallax*; 30: *H. obovatus;*	1,16,20: stem bark; 1,4,7,16,30,31:root; 1,3: wood; 10,19,23,29:stem; 11,16:leaves; 16,24:bark; 16:latex	1: [42]; 1: [48]; 1: [58]; 1: [74]; 1: [75]; 1: [84]; 1,3: [85]; 1,4,7: [38]; 1,4,7: [43]; 4,7: [84]; 10: [39]; 11: [49]; 11,29,30,31: [89]; 16: [51]; 16: [88], 16: [90]; 16: [91]; 16: [92]; 19: [45]; 19: [46]; 20: [86]; 23: [47]; 24: [87]; 30,31: [93]; 30,31: [94]
isoplumieride	1: *Plumeria rubra;*3: *P. acutifolia;* 16: *H. sucuuba;* 27: *H. bracteatus;* 28: *H. stenophyllus*	1,3:root; 16,27,28:leaves; 16,27,28:bark; 16,27,28:latex	1: [65]; 3: [38]; 3: [40]; 3: [43]; 3: [66]; 16: [51]; 16: [73]; 16,27,28: [95]
loganic acid	25: *Rauwolfia grandiflora;* 32: *Catharanthus roseus*	25: bark; 32:seed	25: [1]; 32: [4]
loganin	12: *Alstonia boonei;*15: *A. scholaris;* 13: *Cerbera manghas;*17: *Vinca* sp.*;* 25: *Rauwolfia grandiflora;* 32: *Catharanthus roseus*; 35: *Winchia calophylla*	12,13,17:leaves; 15,25:bark; 32:seed; 35:stem bark	12: [64]; 13: [46]; 13: [56]; 13; [61]; 13: [62]; 15: [60]; 17: [42]; 17: [96]; 25: [1]; 32: [4]; 35: [129]
3-*O*-methylallamancin	10: *Allamanda neriifolia*	10:leaves	10: [39]; 10: [40]
3-*O*-methylallamcim	10: *Allamanda neriifolia*	10:leaves	10: [39]; 10: [40]
naresuanoside	34: *Alstonia macrophylla*	34:stem	34: [128]
obtusadoid A	7: *Plumeria obtusa*	7:aerial parts	7: [72]
obtusadoid B	7: *Plumeria obtusa*	7:aerial parts	7: [72]
plumenoside (β-dihydroplumericinic acid glucosylester)	1: *Plumeria rubra;* 3: *P. acutifolia*	1,3:root	1: [65]; 3: [38]; 3: [40]; 3: [43]; 3: [66]
plumeric acid 1-*β*-*O*-*β*-D-glucopyranosyl	24: *Himatanthus articulata*	24:bark	24: [87]
plumericidine	1: *Plumeria rubra*	1: flowers	1: [97]
plumericin	1: *Plumeria. rubra*, 3: *P. acutifolia;* 4: *P. alba*, 5: *P. bracteata*, 7: *P. obtusa*,	1,16,20:stem bark; 1,4,7,8,9,16,33:root; 1,3:wood; 5,8,16:root cortex; 1,19,23,29:stem;	1: [42]; 1: [48]; 1: [58]; 1: [74]; 1: [75]; 1: [82]; 1: [98]; 1,3: [85]; 1, 4, 7: [43]; 1,4,5,7,8,9: [38];

IRIDOIDS	SPECIES	PLANT MATERIAL	REF.
	8: *P. rubra* var. *alba*, 9: *P. multiflora;*20: P. bicolor; 21: *P. acuminata*;10: *Allamanda neriifolia*;11: *A. cathartica*; 23: *A. schottii*; 31: *A. doniana*;16: *Himatanthus sucuuba*;19: *H. phagedaenicus*; 24: *H. articulatus*; 29: *H. fallax*; 30: *H. obovatus*; 33: *Nerium indicum*	1,10,11,16,23:leaves; 1,16,20,24:bark; 1:fruits; 16:latex	1,21: [44]; 4,7: [84]; 5: [81]; 8: [82]; 9: [99]; 10: [39]; 11: [49]; 11,29,30,31: [89]; 16: [50]; 16: [51]; 16: [88]; 16: [90]; 16: [91]; 16: [92]; 16: [100]; 19: [45]; 19: [46]; 20: [86]; 20: [101]; 23: [47]; 23: [102]; 24: [87]; 33: [84]; 33: [130]
plumeridoid A	1: *Plumeria rubra*	1: stem bark	1: [48]
plumeridoid B	1: *Plumeria rubra*	1: stem bark	1: [48]
plumeridoid C	1: *Plumeria rubra*; 16: *Himatanthus sucuuba*	1,16: stem bark	1: [48]; 16: [103]
plumiepoxide	10: *Allamanda neriifolia*	10:stem; 10:leaves	10: [39]; 10: [40]
plumieride	1: *Plumeria rubra* L., 2: *P. lancifolia*, 3: *P. acutifolia*, 4: *P. alba*,5: *P. bracteata*, 6: *P. obtusifolia*, 7: *P. obtusa*,8: *P. rubra var. alba;* 20: *P. bicolor;* 26: *P. inodora;*10: *Allamanda neriifolia;*11: *A. cathartica;* 23: *A. schottii;* 16: *Himatanthus sucuuba;* 24: *H. articulatus;* 27: *H. bracteatus;* 28: *H, stenophyllus;* 29: *H. fallax*	1,2,7,8:stem bark, 1,3,4,8:root, 1,2,3,4,5,7,8,10,11,16,27 ,28:leaves, 1,8:stem wood, 1,2,3,5:wood, 1,4,7:seed, 1,2,3,4,5,7:flowers, 1:descorticated stem, 2,3,5:root cortex, 2,3,5:pith, 2,3,4,5:cortex, 3:rind; 10,20,23,26,29: stem; 16,20,24,27,28: bark; 16,23,27,28: latex	1: [42]; 1: [48]; 1: [65]; 1: [75]; 1: [76]; 1: [77]; 1: [82]; 1: [84]; 1: [105]; 1: [111]; 1,2,3,4,5,6,7,8,: [38]; 1,4,7,20: [86]; 1,3: [85]; 1,8: [104]; 2: [81]; 3: [66]; 3: [80]; 3: [81]; 3: [106]; 3: [107]; 4: [84]; 4: [108]; 4: [109]; 5: [81]; 6: [110]; 7: [84]; 8: [105]; 10: [39]; 11: [49]; 11: [112]; 16: [51]; 16: [73]; 16,27,28: [95]; 20: [114]; 23: [102]; 24: [87]; 24: [115]; 26: [113]; 29: [89]
plumieride coumarate (13-*O*-coumaroylplumieride; plumieride *p*-*E*-coumarate)	1:*Plumeria rubra*; 3: *P acutifolia;* 4: *P. alba*; 7: *P. obtusa;* 10: *Allamanda neriifolia;*11: *A. cathartica*	1:bark; 1,3,4,7,11:root; 1,4,7:stem; 1,4,7,10: leaves; 1,7: flowers; 1,3: wood	1: [44]; 1,3: [85]; 1,3,4,7: [38]; 1,3,4,7: [43]; 1,4,7: [84]; 3: [66]; 7: [37]; 10: [39]; 11: [89]; 11: [116]
plumieride coumarate glucoside (protoplumericin A; 13-*O*-*p*-*O*-glucosylcoumaroyl plumieride)	1: *Plumeria rubra*; 3: *P. acutifolia*; 4: *P. alba*, 7: *P. obtusa;* 10: *Allamanda neriifolia*; 11: *A. cathartica*	1,3,4,11:root; 1,3:wood, 4,7,10:stem; 4,7,10:leaves; 4,7:flowers; 1:bark	1,3: [85]; 1,3,4,7: [38]; 1,4,7: [43]; 1,4,7: [84]; 1,7,11: [118]; 3: [66]; 10: [39]; 10: [40]; 10: [116]; 10: [117]; 11: [89]
plumieride *p*-*Z*-coumarate	7: *Plumeria obtusa*	7:leaves	7: [37]; 7: [38]

IRIDOIDS	SPECIES	PLANT MATERIAL	REF.
plumieride-1α	1: *Plumeria rubra;*3: *P. acutifolia;* 7: *P. obtusa*	1,3:root; 7:leaves; 7:aerial parts	1: [65]; 3: [38]; 3: [40]; 3: [66]; 7: [72]; 7: [119]
plumieride-1*β*-*O*-*β*-D-glucopyranosyl	24: *Himatanthus articulatus*	24:bark	24: [87]
plumieridin A	1: *Plumeria rubra;* 7: *Plumeria obtusa*	1:flowers; 7:aerial parts	1: [43]; 1: [118]; 1: [120]; 7: [72]
plumieridin B	1: *Plumeria rubra*	1: flowers	1: [43]; 1: [120]
protoplumericin A-1α	1: *Plumeria rubra*; 3: *P. acutifolia*; 10: *Allamanda neriifolia*	1, 3:root; 10:stem; 10:leaves	1: [65]; 3: [38]; 3: [40]; 3: [66]; 10: [39]
protoplumericin B (13-*O*-*p*-*O*-glucosylcaffeoyl-plumieride)	10: *Allamanda neriifolia*	10:stem; 10:leaves	10: [39]; 10: [40]
pulosarioside	14: *Alyxia reinuardtii*	14:bark	14: [40]; 14: [121]
scholarein A	15: *Alstonia scholaris*	15:bark	15: [60]
scholarein B	15: *Alstonia scholaris*	15:bark	15: [60]
scholarein C	15: *Alstonia scholaris*	15:bark	15: [60]
scholarein D	15: *Alstonia scholaris*	15:bark	15: [60]
secologanin	32: *Catharanthus roseus*	32:leaves	32: [4]
sweroside	34: *Alstonia macrophylla*	34:stem	34: [128]
theveside	13: *Cerbera manghas*; 18: *Thevetia peruviana*	13:fruit; 13,18:leaves; 18:root	13: [46]; 13: [56]; 13: [61]; 13: [124]; 13,18: [42]; 13,18: [62]; 18: [122]; 18: [123]; 18: [125]
theviridoside	13: *Cerbera manghas*; 18: *Thevetia peruviana*	13:fruit; 13:cortex; 13,18:leaves; 18:root	13: [46]; 13: [56]; 13: [61]; 13: [68]; 13: [124]; 13,18: [42]; 13,18: [62]; 18: [125]; 18: [126]
theviridoside-10-*O*-*β*-D-fructofuranosyl	18: *Thevetia peruviana*	18:leaves; 18:root	18: [125]; 18: [127]
theviridoside-10-*O*-*β*-D-glucopyranosyl	18: *Thevetia peruviana*	18:root	18: [125]
theviridoside-3'-*O*-*β*-D-glucopyranosyl	18: *Thevetia peruviana*	18:root	18: [125]
theviridoside-6'-*O*-*β*-D-glucopyranosyl	18: *Thevetia peruviana*	18:leaves; 18:root	18: [125]; 18: [127]

Table 1. Iridoids found in Apocynaceae family.

http://dx.doi.org/10.5772/55784

2. Extracts preparation and iridoids isolated

Table 2 illustrates the extraction methods and the iridoids isolated of some genera of Apocynaceae.

SPECIES / PLANT MATERIAL	SOLVENT/ EXTRACTION METHOD	ISOLATED IRIDOIDS	REF.
Allamanda neriifolia Hook./ stem and leaves	MeOH/ Percolation	Isoallamandicin, allamcin, 3-*O*-methylallamcin, allamancin, 3-*O*-methylallamancin, allamcidin, plumericin, isoplumericin, allamandin, allamandicin, plumieride 13-*O*-acetate, deglucosylplumieride, 13-*O*-coumaroylplumieride, allamcidin B β-D-glucoside, plumiepoxide, protoplumericin B, plumieride, protoplumericin A, gardenoside, 10-dehydrogardenoside	[39]
A. neriifolia Hook./ aerial parts	EtOH/ Maceration	Allamanoid, plumieride, protoplumericin, nicotiflorin	[52]
A. neriifolia Hook./ stem	MeOH/ Percolation	Plumericin, plumieride, protoplumericin	[117]
A. schottii Pohl/ stem	EtOH/ Maceration	Allamandin, allamcin, plumericin, isoplumericin, scoparone, scopoletin, pinoresinol	[47]
A. cathartica L. / root bark	EtOH/ Hot extraction	Allamandin, allamandicin, allamdin, plumericin, isoplumericin	[41]
A. cathartica L. /root	MeOH/ Hot extraction	Isoplumericin, plumericin, plumieride, plumieride coumarate, plumieride coumarate glucoside	[116]
A. cathartica L. / root bark	MeOH and $CHCl_3$/ Not described	Plumericin, isoplumericin, plumieride, plumieride coumarate, plumieride coumarate glucoside	[84]
A. cathartica L./ root (inner part)	MeOH and $CHCl_3$/ Soxhlet apparatus	Plumieride, plumieride coumarate, plumieride coumarate glucoside	[84]
A. cathartica L. / leaves	EtOAc/ Not described	Isoplumericin, plumericin	[89]
Alstonia macrophylla Wall. ex G. Don / stem	EtOH/ Maceration	Sweroside, naresuanoside	[128]
A. scholaris (L.) R. Br./ bark	EtOH/ Maceration	Scholarein A, scholarein B, scholarein C, scholarein D, isoboonein, alyxialactone, loganin	[60]

SPECIES / PLANT MATERIAL	SOLVENT/ EXTRACTION METHOD	ISOLATED IRIDOIDS	REF.
Alyxia reinwardtii Blume/ leaves	MeOH/ Maceration	Alyxialactone, 4-epi-alyxialactone	[59]
Cerbera manghas L./ leaves	MeOH/ Not described	Theviridoside, theveside	[124]
C. manghas L./ stem bark and root bark	MeOH/ Percolation	Cerbinal, cerberic acid, cerberinic acid	[67]
C. manghas L./ leaves	MeOH/ Percolation	Cerberidol, epoxycerberidol, cyclocerberidol, cerberidol-3-*O*-β-D-allopyranoside, cerberidol-3,10-bis-*O*-β-D-allopyranoside, epoxycerberidol-3-*O*-β-D-allopyranoside, cyclocerberidol-3-*O*-β-D-allopyranoside	[55]
C. manghas L./ leaves	MeOH/ Percolation	Theviridoside, 10-carboxyloganin, loganin, cyclocerberidol-3-*O*-β-D-glucoside, epoxycerberidol-3-*O*-β-D-glucoside	[62]
C. manghas L./ leaves	MeOH/ Percolation	10-*O*-benzoyltheveside, 10-dehydrogeniposide, loganin, theviridoside, theveside	[61]
C. odollam Gaertn./ leaves	MeOH/ Percolation	Cerberidol, cyclocerberidol, cerberidol-3-O-β-D-allopyranoside, cyclocerberidol-3-O-β-D-allopyranoside	[55]
Himatanthus articulatus (Vahl) Woodson/ bark	MeOH/ Maceration	1β-*O*- β -D-glucopyranosylplumeric acid, plumeride-1 β -*O*- β -D-glucopyranosyl, plumericin, isoplumericin	[87]
H. fallax (Müll. Arg.) Plumel/ stem	EtOAc/ Maceration	Isoplumericin, plumericin, plumieride	[89]
H. sucuuba (Spruce ex Müll. Arg.) Woodson / latex	H_2O/ Not described	15-demethylisoplumieride, 15-demethylplumieride, plumieride, isoplumieride	[70]
H. sucuuba (Spruce ex Müll. Arg.) Woodson/ bark	EtOH/ Maceration	Isoplumericin, plumericin	[88]
H. sucuuba (Spruce ex Müll. Arg.) Woodson / latex	H_2O/ Not described	Plumericin, isoplumericin	[92]
H. sucuuba (Spruce ex Müll. Arg.) Woodson /	BuOH/ Partition	Plumieride, isoplumieride, 15-demethylplumieride	[73]

SPECIES / PLANT MATERIAL	SOLVENT/ EXTRACTION METHOD	ISOLATED IRIDOIDS	REF.
latex			
H. sucuuba, (Spruce ex Müll. Arg.) Woodson / bark	EtOH/ Maceration	Plumericin	[100]
Nerium indicum Mill. / root	Petrol/ Soxhlet apparatus	Plumericin	[130]
Plumeria acutifolia Poir. / root	MeOH/ Percolation	Plumericin, 13-*O*-coumaroylplumieride, 13-O-caffeoylplumieride, 13-deoxyplumieride, plumenoside, 1 α-plumieride, 1 α-protoplumericin A, plumieride, protoplumericin A, 8-isoplumieride	[66]
P. acutifolia Poir. / leaves	MeOH/ Exhaustive Maceration	15-demethylplumieride	[71]
P. bicolor Seem. / bark	MeOH/ Exhaustive Maceration	Plumericin, isoplumericin	[86]
P. rubra L. var. *acutifolia* (Poir.) Woodson / bark	MeOH/ Maceration	15-demethylisoplumieride, 15-demetilplumieride, plumieride, isoplumieride	[70]
P. rubra L. / steam bark	CH_2Cl_2:MeOH (1:1) and MeOH/ Maceration	Plumericin, isoplumericin, plumieride, fulvoplumierin	[75]
P. rubra L. / heartwood	EtOAc/ Percolation	Plumericin, isoplumericin, plumieride, l3-*O*-coumaroylplumieride , protoplumericine A	[85]
P. rubra L. / stem bark	MeOH/ Maceration	Fulvoplumierin, allamandin, allamcin, plumericin, 15-demethylplumieride, plumieride, α-allamcidin, β-allamcidin, 13-*O*-*trans*-p-coumaroylplumieride	[44]
P. rubra L. / stem bark	MeOH/ Maceration	Fulvoplumerin, dihydroplumericin, plumieride , plumeridoids A, B and C, isoplumericin, plumericin, allamcin, allamandin, mixture of plumeridoid C and epiplumeridoid C	[48]
P. rubra L. var. *acutifolia* (Poir.) Woodson /	EtOH/ Maceration	Plumieridin A, plumieridin B, plumericin, plumieride	[120]

SPECIES / PLANT MATERIAL	SOLVENT/ EXTRACTION METHOD	ISOLATED IRIDOIDS	REF.
flowers			
P. rubra L. var. *acutifolia* (Poir.) Woodson/ flowers	EtOH/ Maceration	Plumericidine	[97]
P. obtusa L. / flower	EtOH/ Maceration	Plumieride coumarate glucoside	[118]
P. obtusa L. / aerial parts	MeOH/ Maceration	Obtusadoids A and B, plumieridin A, plumieridine, 1α-plumieride, 15-demethylplumieride	[72]
P. obtusa L. / leaves	MeOH/ Maceration	6"-*O*-acetylplumieride *p*-E-coumarate, 6"-*O*-acetylplumieride,*p*-Z-coumarate, plumieride, plumieride *p*-Z-coumarate, plumieride *p*-E-coumarate	[37]
P. obtusa L. / leaves	MeOH/ Maceration	1α-plumieride	[119]
P. inodora Jacq. / stems	H_2O/ Maceration	Plumieride	[113]
P. bicolor Seem. / bark	MeOH/ Maceration	Plumieride	[114]
Rauwolfia grandiflora Mart. Ex A. DC. / bark	EtOH/ Maceration	Loganic acid, loganin, boonein, isoboonein	[1]
Thevetia peruviana (Pers.) K. Schum. / Leaves	MeOH/ Percolation	10-*O*- β -D-fructofuranosyltheviridoside, 6'*O*- β -D-glucopyranosyltheviridoside	[125]
T. peruviana (Pers.) K. Schum. / root	MeOH/ Percolation	Theveside, theviridoside, 10-*O*- β -D-fructofuranosyltheviridoside, 6'*O*- β -D-glucopyranosyltheviridoside, 10-*O*- β -D-glucopyranosyltheviridoside, 3'*O*- β -D-glucopyranosyltheviridoside	[127]
Winchia calophylla A. DC. / stem bark	EtOH/ Hot extraction	Loganin	[129]

Table 2. Main extraction methods to obtain iridoids of Apocynaceae family.

http://dx.doi.org/10.5772/55784

3. Chemical structures

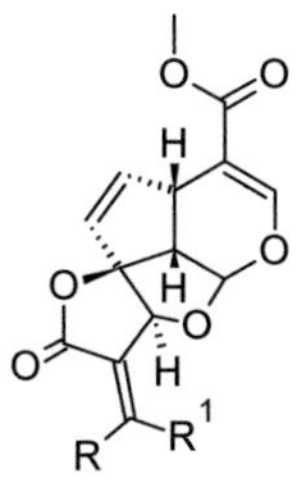

1 *Plumericin* R = H; $R^1 = CH_3$

2 *Isoplumericin* $R = CH_3$; $R^1 = H$

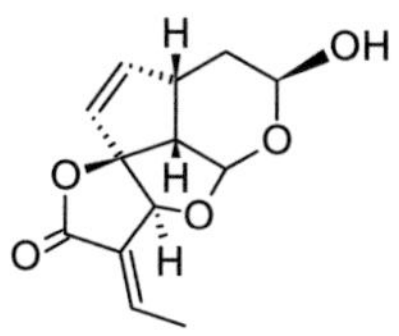

3 *Allamcin*

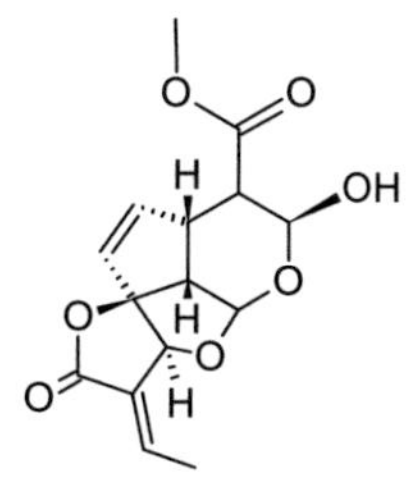

4 *Allamandin*

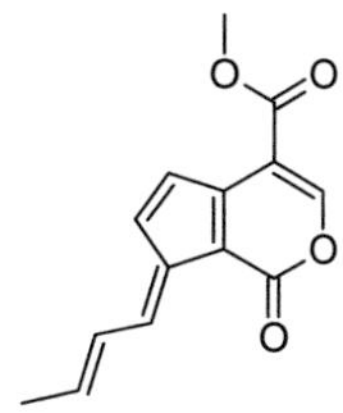

5 *Fulvoplumierin*

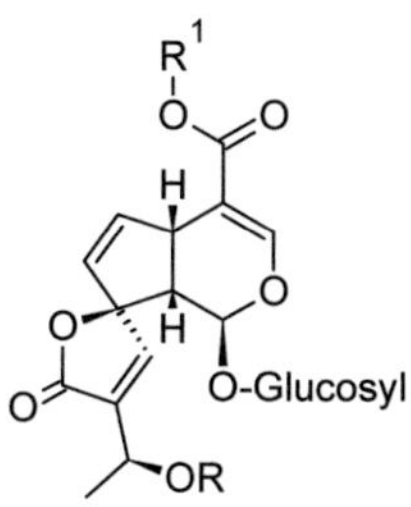

6 *Plumieride* R = H; $R^1 = CH_3$

7 *15-demethylplumieride* $R=R^1= H$

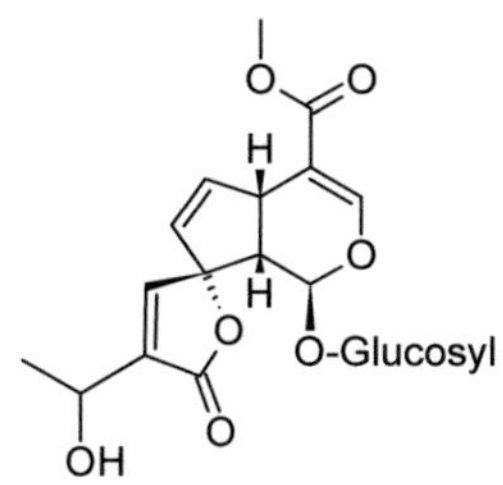

8 *Isoplumieride*

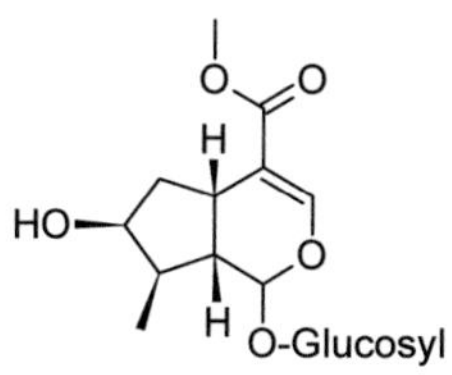

9 *Loganin*

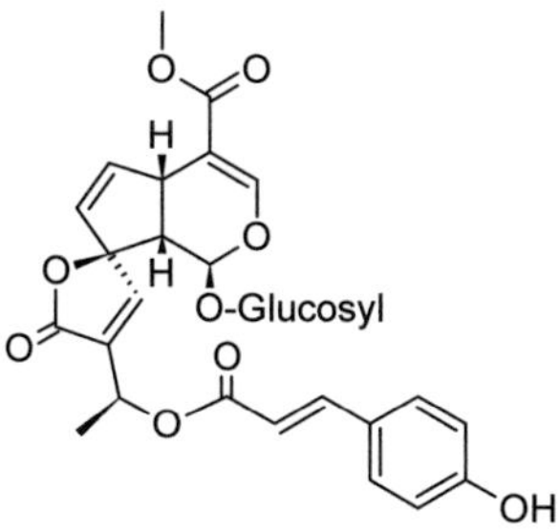

10 *Plumieride coumarate*

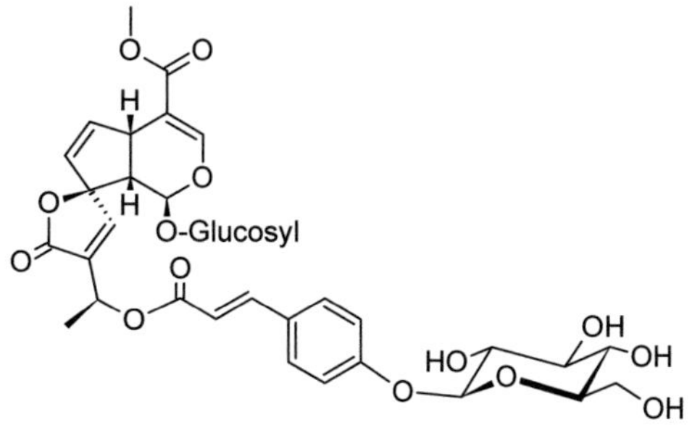

11 *Plumieride coumarate glucoside*

Figure 2. The most isolated iridoids of Apocynaceae family

4. Chromatographic separation

4.1. Thin layer chromatography

Preparative chromatography was performed with thin layer chromatography (TLC) aluminum sheets and 8:2 chloroform/ methanol as mobile phase of a fraction from the 95% EtOH extract of stems of *Alstonia macrophylla*. This procedure led to the isolation of the compounds sweroside (2 mg) and naresuanoside (3 mg) [128]

Another interesting application of preparative chromatography is described in [47] with ethanol extract of ground stems of *A. schottii*. This extract was fractionated by bioassay monitoring and after successive liquid-liquid partition and flash column chromatography; the authors obtained the iridoids allamandin; allamcin and a mixture of plumericin and isoplumericin. This separation was performed with a Chromatotron rotor (silica gel (Si gel), 2-mm) with 1 % methanol in chloroform as solvent system.

The bioassay-guided fractionation of the extracts of *Plumeria rubra* barks also proved to be a successful strategy, leading to the isolation of eleven substances, of which eight were iridoids. In this context, the aqueous extract of *Plumeria rubra* bark has iridoids such as the epimers, α-allamcidin and β-allamcidin, which were resolved by preparative TLC on Si-gel G plates (20 × 20 cm, 250 μm, Merck®), using chloroform/ ethyl acetate/ methanol (3:3:1) as solvent system [44].

Ferreira and coworkers [95] describe the HPTLC analysis of the bark, latex and leaf extracts and substances of *Himatanthus sucuuba*. The solutions of the extracts in methanol (10 mg mL $^{-1}$) and the isolated iridoids (3 mg mL^{-1} - plumieride; 1 mg mL $^{-1}$ - isoplumieride) were applied (5 μL) using Linomat IV, a Camag semi-automatic spotter. The analysis was carried out on a precoated silica gel 60 F254 (Merck) HPTLC plate (0.2 mm of layer thickness and 10 × 10 cm size) using chloroform/ methanol (8:2) as developing system. The resulting chromatogram was dried and the spots were visualized by spraying with vanillin – sulfuric acid solution, followed by heating at 100 °C. Table 3 shows other studies using TLC analysis of Apocynaceae iridoids.

SPECIES	PLANT MATERIAL	IRIDOIDS	STATIONARY PHASE	MOBILE PHASE	REF.
Plumeria obtusa	Flowers	Plumieride coumarate glucoside	Si gel 60G F_{254}	$CHCl_3$-MeOH (3:1)	[118]
Plumeria obtusa	Leaves	6"-*O*-acetylplumieride *p*-E-coumarate	1- Si gel GF_{254}	1- $CHCl_3$-MeOH (8.5:1.5)	[37]
		6"-*O*-acetylplumieride *p*-Z-coumarate	2- Si gel	2- Gradients of Petrol-EtOAc	
Plumeria rubra	Stem bark	Plumeridoid A	1- Si gel GF_{254}	1- Hexane-EtOAc (20%)	[48]
		Plumeridoid B	2- Si gel GF_{254}	2- Hexane-EtOAc (4%)	
Himatanthus fallax	Stems	Plumericin Isoplumericin	Si gel	10% Me_2CO in $CHCl_3$	[89]
		Plumieride		$CHCl_3$-MeOH (85:15)	

SPECIES	PLANT MATERIAL	IRIDOIDS	STATIONARY PHASE	MOBILE PHASE	REF.
Himatanthus sucuuba	Latex	Plumericin	Si gel GF_{254}	Hexane-EtOAc (6:4)	[92]
		Isoplumericin			
Alyxia reinwardti	Leaves	Alyxialactone	Si gel GF_{254}	$CHCl_3$-MeOH (95:5)	[59]
		4-*epi*-alyxialactone			
Allamanda cathartica	Roots	Plumericin	1- Si gel	1- $CHCl_3$	[41]
		Isoplumericin			
		Allamandicin	2- Si gel	2- $CHCl_3$-MeOH (95:5)	
		Allamdin	3- Si gel	3- ether-hexane (1:1)	

Table 3. TLC analyses of Apocynaceae iridoids.

4.2. Open column chromatography

The methanol extract of the leaves of *Cerbera manghas* and its fruit contain the iridoids theveside and theviridoside, as described in [124]. The methanol extract of the leaves, after the addition of water, was sequentially partitioned with chloroform, acetic acid and butanol. This extract and the aqueous phase were submitted to column chromatography with charcoal and water/methanol as eluent. Theveside was isolated from the aqueous phase. Fractions of the butanol extract, which turn blue after heating with mineral acid, were chromatographed over a silica gel column with a gradient of increasing polarity of chloroform/methanol to afford theviridoside.

Cerberidol, epoxycerberidol, cyclocerberidol, cerberidol-3-*O*-*β*-D-allopyranoside, cerberidol-3,10-bis-*O*-*β*-D-allopyranoside, epoxycerberidol-3-*O*-*β*-D-allopyranoside and cyclocerberidol-3-*O*-*β*-D-allopyranoside are present in the leaves of *C. manghas* [55]. The methanol extract from 1.45 kg of dried leaves was concentrated, re-suspended in water, and sequentially partitioned with benzene and chloroform. The aqueous phase was chromatographed using MCI-gel column (elution with gradients of water/methanol) and the fraction eluted with 20% methanol was chromatographed in two steps: using RQ-1 (Fuji-gel phase) column (elution with water/acetonitrile) and using a silica gel column (elution with gradient of chloroform/methanol/water 7:3:1 to 7:3:1.2). Fractionation afforded cerberidol (75 mg), epoxycerberidol (10 mg), cyclocerberidol (160 mg), cerberidol-3-*O*-*β*-D-allopyranoside (135 mg), cerberidol-3,10-bis-*O*-*β*-D-allopyranoside (20 mg), epoxycerberidol-3-*O*-*β*-D-allopyranoside (27 mg) and cyclocerberidol-3-*O*-*β*-D-allopyranoside (250 mg). Some of these iridoids are also found in dried leaves of *Cerbera odollam* Gaertn [55]. The same methodology was used to isolate cerberidol (15 mg), cyclocerberidol (48 mg), cerberidol-3-*O*-*β*-D-allopyranoside (28 mg) and cyclocerberidol-3-*O*-*β*-D-allopyranoside (14 mg).

The iridoids cerbinal, cerberic acid and cerberinic acid are found in the methanol extract of the bark of *C. manghas* [67]. The crude extract obtained by percolation from 4 kg of stem bark and 1.9 kg of root bark were diluted with water to 50% water:methanol. The mixture was washed with hexane and partitioned with benzene. Benzene fractions were re-suspended in methanol and cerbinal (120 mg from stem bark extract and 300 mg from root bark extract) precipitated.

The supernatant was chromatographed on a silica gel column and cerbinal, cerberic acid and cerberinic acid were eluted with benzene/acetone.

The iridoids isoplumericin, plumericin, plumieride, plumieride coumarate and plumieride coumarate glucoside can be detected and quantified in several species of *Plumeria* and *Allamanda* by TLC using silica gel 60 and the following mobile phases: benzene/ethyl acetate 4:1, chloroform/methanol 4:1, chloroform/methanol 7:3, propanol/ethyl acetate/water 7:2:1 [84]. Visualization of the chromatograms is achieved by spraying with 50% sulfuric acid/ ethanol solution and heating. For the analyses, it is necessary to use iridoids as standards that can be isolated from the roots of *Allamanda cathartica*. For the isolation of iridoids during the development of the method, chloroform and methanol extracts, sequentially obtained in a Soxhlet apparatus from the root bark and inner roots of *Allamanda cathartica* (15.0 g for each extract) were successively fractionated on column chromatography. In the first chromatography step, the chloroform extract from the bark (1.7 g) was applied to a silica gel column (80 g) and eluted with gradient of petrol, ethyl ether, chloroform and methanol. This procedure yielded a mixture containing isoplumericin and plumericin (160 mg). Fractions eluted with chloroform/ methanol (3:2) were mixed with the methanol extract of the bark (total mass: 2.8 g) and chromatographed on a silica gel column (150 g deactivated with water) with gradient of chloroform and methanol as mobile phase to furnish plumieride coumarate (150 mg), plumieride coumarate glucoside (480 mg) and two mixtures: one containing plumieride and plumieride coumarate, and the other containing plumieride and plumieride coumarate glucoside. The first mixture was further resolved on partition between water and ethyl acetate and afforded 180 mg of plumieride and 150 mg of plumieride coumarate. The other mixture was rechromatographed to afford plumieride (300 mg) and plumieride coumarate glucoside (410 mg). Neither the chloroform (810 mg) nor the methanol (920 mg) extracts from the inner part of the roots contained isoplumericin and plumericin. These extracts were purified on silica gel column (75 g) deactivated with water and eluted with chloroform and methanol gradient, to give plumieride (70 mg), plumieride coumarate (80 mg) and plumieride coumarate glucoside (200 mg).

The fresh leaves of *Cerbera manghas* contain the iridoids 10-*O*-benzoyltheveside, 10-dehydrogeniposide, loganin, theviridoside and theveside [124]. For the isolation, the methanol extract obtained by percolation using 2.6 kg of fresh leaves was extracted with butanol, and this extract was partitioned with benzene. After partition, the remaining butanol fraction (22.8 g) was subjected to column chromatography using MCI-gel (CHP-20) as stationary phase and a gradient of methanol/water as eluent. The fraction eluted with 20% methanol was subjected to C-18 column (elution with acetonitrile/water) to afford 20 mg of 10-dehidrogeniposide and 20 mg of loganin. The fraction eluted with pure water (2.6g) in the first chromatographic step was also subjected to C-18 column (elution with acetonitrile/ water) to furnish 23 mg of 10-*O*-benzoyltheveside and 270 mg of theveside.

According to [44] the iridoids fulvoplumierin, allamandin, α- and β-allamcidin, plumieride, 15-demethylplumieride, 13-*O*-*trans*-*p*-coumaroylplumieride and plumericin are present in extracts of *Plumeria rubra*. For isolation of these iridoids, the stem bark (2.5 kg) was successively extracted with petroleum ether and methanol followed by bioguided fractionation to investigate the cytotoxic activity against various cancer cells, particularly murine lymphocytic leukemia cells (P-388). Petroleum ether extract (30 g) was submitted to flash column chromatography over silica gel (750 g, 230-400 mesh) with chloroform/petroleum ether (1:1) as eluent. Fraction 002 (200 mg) was rechromatographed using silica gel (120 g) and chloroform/methanol (99:1) to isolate fulvoplumierin (25 mg after recrystallization from petroleum ether/chloroform 1:1). The methanol extract (295 g) was partitioned between chloroform and water. The chloroform extract (60 g) was successively chromatographed in a silica gel column (1.5 kg) with gradients of increasing polarity, with chloroform and methanol, and then in a silica gel (600 g) with petroleum ether/chloroform/ethyl acetate (1:3:1) as eluent to furnish allamandin (12 mg after recrystallization from chloroform). Another fraction from the first column chromatography of the methanol extract (1.2 g) was purified in a silica gel column (400 g) and ethyl acetate/ methanol (97:3) and recrystallyzed from chloroform to furnish 9 mg of allamcin. The aqueous extract (200 g) was also subjected to column chromatography using silica gel column (2 kg) and gradient of chloroform and methanol. Plumieride (55 g) was obtained directly from fraction F 022 after recrystallization from methanol. Fraction F022 (29 g) also furnished 15-demethylplumieride and 13-*O*-*trans*-*p*-coumaroylplumieride (3 g) by column chromatography over silica gel and gradient of chloroform and methanol. Fraction F018 (800 mg) was purified over silica gel (400 g) with gradient of ethyl acetate and methanol to afford plumericin (18 mg after recrystallization with ethyl acetate). Another fraction (F019) was purified over silica gel (250 g) with ethyl acetate/chloroform/methanol (6:6:1) solvent system to furnish an unstable iridoid aldehyde and a mixture of α- and β-allamcidin (12 mg and 16 mg, respectively) further isolated by TLC on silica gel plates (20 cm x 20 cm, 250 μm) with ethyl acetate/chloroform/methanol (3:3:1).

Plumericin, isoplumericin, plumieride, 13-*O*-coumaroylplumieride and protoplumericine A were isolated from the ethyl acetate extract of *P. rubra* heartwood [85]. The extract (44 g) obtained by percolation was submitted to column chromatography on silica gel with light petroleum ether/ethyl acetate (1:1) and methanol as eluent to furnish four fractions (A-D). Fraction C directly afforded plumericin (1.22 g) after recrystallization from toluene/ethyl acetate. Fraction B was rechromatographed on a silica gel with toluene/ethyl acetate (9:1), and isoplumericin (140 mg) was isolated from fraction 2. Fraction D was also rechromatographed on a silica gel column, but with chloroform/methanol/water (90:10:0.5 to 70:30:10) as solvent system. Fraction 3′ was 13-*O*-coumaroylplumieride (2.0 g) and fraction 7′ was protoplumericine (1.6 g). Fraction 5′ was subjected to Sephadex LH20 column eluted with methanol, followed by silica gel column eluted with chloroform/ methanol/water (85:15:0.7) to furnish plumieride (120 mg).

The iridoids, plumericin, isoplumericin, plumieride and fulvoplumierin, were present in the extracts of *Plumeria rubra* bark. After maceration of the powdered bark (3.5 kg) with dichloromethane/methanol (1:1) and pure methanol, the combined extracts were partitioned between water and ethyl acetate. To isolate the four iridoids, the organic layer was chromatographed twice in a column using silica gel and gradient of increasing polarity with hexane and ethyl acetate, ethyl acetate and methanol, and then pure methanol. The amounts of the compounds isolated were not reported [75].

The flowers of *Plumeria rubra* L. cv. acutifolia can provide plumericidine, as described by [97]. The ethanol (95%) extract, obtained from 2.9 kg of flowers, was successively partitioned with petroleum ether, ethyl acetate and butanol. The ethyl acetate fraction was sequentially submitted twice to column chromatography using silica gel and gradient of chloroform/ methanol as mobile phase. Chromatography on a Sephadex LH20 column yielded 20 mg of plumericidine.

According to [72], several iridoids can be isolated from the aerial parts of *Plumeria obtusa*: obtusadoid A, obtusadoid B, plumieridin A, 1α-plumieride, 15-demethylplumieride and plumieridine. The methanol extract (400 g) was obtained from 10 kg of the plant material and sequentially partitioned with hexane and ethyl acetate. The ethyl acetate extract was chromatographed using a silica gel column and gradients of hexane, ethyl acetate and methanol. The less polar fractions were rechromatographed in the same stationary phase, and eluted with hexane/dichloromethane (1:1) to afford obtusadoid A (6 mg), obtusadoid B (11.5 mg), plumieridin A (8 mg) and plumieridine (12 mg). The more polar fraction obtained in the first chromatography was filtered on a Sephadex LH20 column using methanol, and further submitted to RP-8 flash column chromatography. Elution with 50% methanol afforded 1α-plumieride (22 mg) and 15-demethylplumieride (13 mg).

Plumieride also can be isolated from the bark of *Plumeria bicolor* [114]. Powdered bark (4 kg) was extracted in methanol, and the crude extract was washed with acetonitrile. The material was re-extracted with chloroform, and this extract was fractionated in column chromatography using silica gel (900 g) and different solvents of increasing polarity. Plumieride was eluted with chloroform/ethyl acetate (1:1) and recrystallized from methanol.

The bark of *Plumeria bicolor* also contains plumericin and isoplumericin, as described in [86]. The methanol extract (100 g), after washing with acetonitrile, was extracted with chloroform and chromatographed on a column containing 800 g of silica gel G (60-120 mesh). Elution was carried out using gradients of increasing polarities with benzene, chloroform and methanol. Plumericin and isoplumericin was recrystallized from methanol.

Isoplumericin and plumericin are present in the bark of *Himatanthus sucuuba* [88]. For the isolation of these iridoids, 95% ethanol extract (2 g), obtained from 50 g of plant material was submitted to column chromatography using silica gel and gradients of increasing polarities with hexane, ethyl acetate and methanol. After recrystallization, isoplumericin (18 mg) was obtained from ethyl acetate and plumericin (70 mg) from methanol.

The stem bark of *Winchia calophylla* contains loganin (1.25 g) [129]. The 95% ethanol extract (600 g) from the dried stem bark (10.5 kg) was partitioned between petroleum ether and water. The petroleum ether extract was submitted to acid-base extraction and after adjustment to pH 9-10 with ammonium hydroxide; the aqueous layer was extracted with petroleum ether, chloroform and butanol. The chromatography of the butanol fraction using silica gel H column led to the isolation of loganin.

The iridoids, scholarein A, B, C and D, can be obtained by the fractionation of the ethanol extract from bark of *Alstonia scholaris* [60]. The crude extract, obtained from 15 kg of the plant material, was partitioned between ethyl acetate and water. The organic layer (190 g) was sequentially chromatographed on a column. The first chromatography, over silica gel (2.1 kg) and using gradient of chloroform and ethyl ether, furnished five fractions (1-5). From fraction 2, 8 mg of scholarein B and 7 mg of scholarein D were isolated after silica gel chromatography and elution with petroleum ether/ethyl ether (3:1). From fraction 3, 25 mg of scholarein A and 60 mg of scholarein C were obtained after successive columns using silica gel and chloroform/ ethyl ether as stationary and mobile phases, respectively.

4.3. Gas Chromatography (GC)

In the study on iridoids, the technique of gas chromatography is generally used for analytical purposes. Gas chromatography represents an advantage over thin layer chromatography, particularly for detecting substances in small amounts, and mass spectrometry can be used to distinguish most iridoid and secoiridoid glucosides by fragmentation patterns [112].

Methods have been developed by [112] for the detection of 33 iridoids and secoiridoid glucosides in mixtures and plant extracts using gas chromatography sometimes coupled to mass spectrometry. For the gas chromatography analyses, a Shimadzu Model GC-1C gas chromatograph with hydrogen FID-1B flame ionization detector was used. Columns were packed with 1.5% OV-1, 1.5% OV-17, 2% OV-210 and 2% OV-225 on 80-100-mesh Shimalite W AW/DMCS. In all, 33 iridoids and secoiridoids glucosides are analyzed as TMS-derivatives. Using the 1.5% OV-17 column with 1.8 m in length and 4 mm in I.D. at 270 ºC, the elution order was: aucubin, 7-deoxyloganic acid and catalpol (retention time = 1.37 min), 7-deoxyloganin, monotropein, gardenoside, secologanin, loganin, scandoside, theviridoside, geniposide, scandoside methyl ester, 7-dehydrologanin, morroniside, hastatoside and forsythide (retention time = 2.55 min), forsythide 10-methyl ester, verbenalin, sweroside, gentiopicroside and swertiamarin (retention time = 3.08 min), bankakosin, kingiside, amaroswerin, amarogentin and asperuloside. The 1.5% OV-17 column, 0.5 m in length and 3 mm in I.D., at 230ºC, furnished the same elution order as above, but forsythide 10-methyl ester, verbenalin and sweroside eluted together (retention time = 3.20 min), while the separation of hastatoside and forsythide, gentiopicroside and swertiamarin were better. When the non-polar 1.5% OV-1 column with 1.8 m in length and 4 mm in I.D. was used at 270 ºC, better separation between loganin and secologanin occurred. Better results were achieved for the 7-deoxyloganic, 7-deoxyloganic acid and catalpol. However, verbenalin and sweroside eluted together, and amaroswerin and amarogentin were not detected. The OV-17 column was slightly polar and, in general, it influenced the larger range of the retention times. The more polar columns with 2% OV-210 at 215 ºC, and 2% OV-225 at 230 ºC, both with 0.5 m in length, showed important differences from keto compounds, such as 7-dehydrologanin and verbenalin, and lactonic compounds, such as sweroside, gentiopicroside and gentiopicroside, reflected in their longer retention times. Amaroswerin and amarogentin were not detected. Sweroside and gentiopicroside were well-separated on OV-210 column, which was not observed using other columns. Paederoside, ligustroside, catalposide, oleuropein, 10-acetoxyligustroside and 10-acetoxyoleuropein were only detected and well-separated in OV-17 and OV-1 columns with 0.5m in length at 270 °C.

For the gas chromatography-mass spectrometry studies, a Hitachi K-53 gas chromatograph and a Hitachi RMU-6 E mass spectrometer were used. The glass columns, 0.5 m x 3 mm in I.D., were packed with 1.5% OV-17 on 80-100-mesh Shimalite W AW/DMCS and 1.5 % OV-1 on 80-100-mesh Shimalite W AW/DMCS, and were used to the oleuropein-type glucosides detection. The authors considered the GC-MS identification of some iridoids was not satisfactory: any important peak different from the sugar moiety was detected in asperuloside and paederoside TMS-derivatives of amaroswerin, amarogaentin asperuloside and paederoside exhibited different retention times, but the same fragmentation pattern; separation using this technique was unsuccessful [112].

Aqueous extracts of different plant species with known presence of iridoid and secoridoid glucosides were analyzed by GC-FID and GC-MS [112]. One of these species was *Allamanda cathartica* var. schottii (Pohl) Rafill (Apocynaceae) cultivated in a greenhouse. Aqueous extracts obtained from 3-5 g of fresh plant material and hot water were treated in a column of charcoal (active carbon for column chromatography) for removal of sugars by elution with water. The sample was eluted with methanol and concentrated under reduced pressure. TMS-derivatives were prepared. For the analyses, a 1.5% OV-17 column with 1.8 m in length at 280 ºC was used. GC-FID and GC-MS (70 eV) showed the presence of one iridoid glycoside. The fragmentation pattern indicated that the original glucoside was plumieride, and that relative retention time was the same as that of asperuloside.

Isoplumericin, plumericin, plumieride, plumieride coumarate and plumieride coumarate glucoside can be detected by GC-FID [116]. For the development of the method, it was necessary to isolate these iridoids for use as standard. The methanol extract of *Allamanda cathartica* L. roots, obtained with 500 g of the dried plant material and boiling methanol, was submitted to silica gel 60 column (2.5 kg) with the eluents: petrol, petrol/ethyl ether, ethyl ether, ethyl ether/chloroform, chloroform/ methanol and pure methanol [116]. Fractionation was monitored by TLC [84]. The fractions were eluted with petrol/ethyl ether until chloroform/ methanol contained isoplumericin and plumericin (4.2 g). This mixture (1.0 g) was suspended in ethyl ether and rechromatographed on silica gel (40 g). Elution with petrol and a gradient of increasing polarity with petrol/ethyl ether furnished isoplumericin (250 mg), plumericin (140 mg) and a mixture of both (290 mg). A second fraction eluted with chloroform/methanol in the first chromatographic step (8.0 g) was resubmitted to column chromatography on silica gel (300 g). Elution with a gradient of chloroform and methanol led to the isolation of plumieride coumarate (5.1 g). The third fraction of the first chromatograph step (10.0 g), eluted with chloroform/methanol, was partitioned between water and ethyl acetate to give plumieride coumarate in the organic phase (3.8 g) and plumieride in the aqueous phase (1.8 g). Finally, 30.0 g of the forth fraction of the first chromatograph step, eluted with chloroform/methanol to methanol, was submitted to column chromatography on silica gel (1.0 kg) deactivated with water. The elution was carried out with a gradient of chloroform and methanol and furnished plumieride coumarate glucoside (6.2 g). GC-FID analyses were used to evaluate the pure grade of fractions and isolated substances. For these analyses, trimethylsilylation of iridoids using HMDS-TMCS and pyridine, and acetylations with acetic anidride and pyridine, were necessary. Plumieride coumarate (isomer mixture) and plumieride coumarate glucoside were acetylated and their products were purified on chromatography with silica gel column and ethyl ether as eluents, to afford pure penta-acetylplumieride coumarate and octa-acetylplumieride coumarate glucoside, respectively. Furthermore, plumieride, plumieride coumarate and plumieride glucoside were hydrolyzed under heating and acid conditions (sulfuric acid, 1 N for 2-3 h), extracted with ethyl acetate and both organic phase and aqueous phase (after neutralization with Amberlite) were analyzed by TLC and GC. Analyses were performed on a 1.5% OV-17 glass column with 0.4 m length and 4 mm I.D. Other analytical conditions were: nitrogen as carrier gas at 50 mL/min; detector temperature, 320 ºC; column temperatures: 190 ºC for isoplumericin and plumericin, 240 ºC for TMS-plumieride, 300 ºC TMS-plumieride coumarate. Glucose was detected in aqueous phases from acid hydrolyses of plumieride, plumieride coumarate and plumieride coumarate glucoside, while *p*-coumaric acid was detected in the organic phases of plumieride coumarate and plumieride coumarate glucoside. GC analyses also showed that plumieride coumarate was isolated as an isomer mixture (20% cis and 80% trans isomers) [116].

4.4. High, Medium and Low Performance Liquid Chromatography (HPLC, MPLC and LPLC)

Studies on HPLC with iridoids of Apocynaceae focus mainly on the separation of components from extracts or fraction. The chromatography profile, the identification and quantification of these terpenes in the extracts are described. Table 4 shows the principal references on iridoids isolated from Apocynaceae by HPLC, MPLC and LPLC.

TECHNIQUE	MOBILE PHASE	COLUMN	SPECIES/ PLANT MATERIAL	SAMPLE	IRIDOID	REF
HPLC	50% methanol in water	µBondapack C-18	*Rauwolfia grandiflora*/ bark	ethanol extract	boonein and isoboonein	[1]
HPLC	10% acetonitrile	Octadecylsilane (ODS)	*Thevetia peruviana*/ root	methanol extract	theviridoside, 10-*O*-*β*-D-fructofuranosyltheviridoside, 6'-*O*-*β*-D-glucopyranosyltheviridoside, 10-*O*-*β*-D-glucopyranosyltheviridoside and 3'-*O*-*β*-D-glucopyranosyltheviridoside	[125]
HPLC	10% acetonitrile	Octadecylsilane (ODS)	*Thevetia peruviana*/ leaves	methanol extract	10-*O*-*β*-D-fructofuranosyltheviridoside and 6'-*O*-*β*-D-glucopyranosyltheviridoside	[127]
HPLC	20 mM KH_2PO_4 and acetonitrile Flow: 0.8 mL/min	Cosmosil 5 C_{18}-AR (5 µm, 25 cm x 4.6 mm I.D.)	*Rauwolfia* / barks	ethanol extract	loganin, loganic acid and gardenoside	[131]

TECHNIQUE	MOBILE PHASE	COLUMN	SPECIES/ PLANT MATERIAL	SAMPLE	IRIDOID	REF
HPLC	water/ acetonitrile	Octadecyl-silane (ODS) (250×10mm)	*Plumeria acutifolia* / leaves	methanol extract	15-demethylplumieride	[71]
HPLC	not informed	Zorbax SB-C18	*Plumeria rubra* L. cv. acutifolia/ flowers	ethyl acetate fraction	plumieridin A and plumieridin B	[120]
HPLC	not informed	not informed	*Allamanda neriifolia*/ aerial parts	ethanol extract	allamanoid, plumieride and protoplumericin	[52]
LPLC	5% ethyl acetate in chloroform	Michel-Miller column (200 g Silica gel)	*Allamanda schottii*/ stem	ethanol extract	allamandin and allamcin	[47]
MPLC	water to 50% methanol in water	RP-18 (25–40 mm particle size, 460 × 36 mm I.D., 460 × 15 mm I.D.)	*Himatanthus sucuuba*/ latex	methanol/ water (1:1)	plumieride	[73]
TECHNIQUE	MOBILE PHASE	COLUMN	SPECIES/ PLANT MATERIAL	SAMPLE	IRIDOID	REF.
HPLC	Acetonitrile and water containing 0.05% trifluoro-acetic acid Flow: 1 mL/min	Lichrospher C18 (5mm, 250 mm × 4.6 mm I.D.)	*Himatanthus sucuuba*/ bark and latex	methanol/ water (1:1)	plumieride, isoplumieride and demethylplumieride	[73]
MPLC and HPLC	water to 50% methanol; 30% acetonitrile in water to 50% acetonitrile MPLC Flow: 6 mL/min	LiChroprep C-18 (45 × 3.5 cm) and Shimpack C-18 (10 μm, 45 cm × 250 mm I.D.)	*Himatanthus sucuuba*/ latex	aqueous fraction and fractions from this procedure	15-demethylisoplumieride, 15-demethylplumieride, plumieride and isoplumieride	[70]

Table 4. Separation by Pressure Liquid Chromatography

4.5. Counterflow

Protoplumericin and plumieride can be extracted from the methanol extract of *Allamanda neriifolina* stems [117] by droplet counter-current chromatography. Crude extract, obtained from 1.6 kg of plant material, was successively partitioned with benzene, chloroform and butanol. Plumericin (420 mg) was directly obtained from the benzene fraction (0.02% yield). The butanol fraction was subjected to sequential chromatographic steps, using XAD-2 column and gradient of water/methanol (mobile phase), silica gel column and solvent system containing chloroform/methanol/water or chloroform/methanol, followed by droplet current chromatography with chloroform/methanol/ water. Plumieride (1.3 g) and protoplumericin (13.2 g) were obtained with yields of 0.08% and 0.83%, respectively.

The iridoids allamcidin B β-D-glucoside, plumiepoxide and protoplumericin B were isolated from *Allamanda neriifolia* extract obtained by percolation with methanol, also by droplet counter-current chromatography [39]. The crude methanol extracts from 2.6 kg of stem and 6.7 kg of leaves were sequentially fractionated with benzene, chloroform and butanol. Previous chromatographic treatment with MCI gel (elution with water/methanol), silica gel column (mobile phases: chloroform/methanol/water; benzene/acetone; ethyl acetate/methanol/ water and ethyl acetate/hexane) and Sephadex LH20 column (mobile phase: chloroform/methanol) led to the isolation of isoallamandicin (10 mg from the stem), allamcin (230 mg from the leaves), 3-*O*-methylallamcin (30 mg from the leaves), allamancin (102 mg from the stem), 3-*O*-methylallamancin (41 mg from the leaves), allamcidin (125 mg from the leaves), plumieride 13-*O*-acetate (760 mg from the stem and *ca.* 2 g from the leaves). Fractions of butanol extracts from the stem and leaves were subjected to droplet counter-current chromatography using chloroform/methanol/water (5:6:4, ascending mode) to obtain allamcidin B β-D-glucoside (17 mg from the stem), plumiepoxide (7 mg from the stem and 374 mg from the leaves) and protoplumericin B (70 mg from the leaves).

Iridoids can be isolated from *Plumeria acutifolia* roots following a similar methodology [66]. Successive liquid-liquid partitions of the crude methanol extract (obtained from 6 kg of plant material) with benzene, chloroform and butanol, followed by several chromatographic steps for fractionation of the butanol fraction (using polystyrene, silica gel and octadecyl silica columns) led to the isolation of 13-*O*-coumaroylplumieride (43 g), plumieride (7.5 g), 13-O-caffeoylplumieride (60 mg), 1α-plumieride (20 mg) and protoplumericin A (9 g). A further purification step involving droplet counter-current chromatography and a mixture of chloroform/methanol/water (4:6:5, ascending mode) led to the isolation of 13-deoxyplumieride (200mg), plumenoside (50 mg) and 8-isoplumieride (700 mg).

4.6. Capillary electrophoresis

For analytical purposes, iridoids can be analyzed by capillary electrophoresis. A method to separate nine iridoids described in [131] uses a Hewlett-Packard (HP^{3D} CE) capillary electrophoresis system coupled to a photodiode array detector (210 nm and 230 nm) and equipped with a fused-silica capillary tube (90 cm × 75μm I.D.). The distance to the detector was 81.5 cm. Other conditions: sample injection at 50 mbar for 3 s and further deionized water injection at 50 mbar for 3 s; constant voltage, 16 kV (positive to negative); cartridge temperature, 30 °C; electrolyte (buffer), 50 mM sodium borate and 30 mg/mL 2,6-di-*O*-methyl-β-cyclodextrin (DM-β-CD); run time, 32 min. Before the analyses, the capillary column was sequentially purged with 0.5 M NaOH, 0.1 M NaOH, deionized water and buffer solution. The iridoids studied eluted in the following order: geniposide, loganin, shanzhiside, aucubin, catalpol, harpagoside, gardenoside, geniposidic acid and loganic acid. All were commercially purchased and only loganin, loganic acid and gardenoside were described for the Apocynaceae family. Several conditions of analyses were studied, including different pH, surfactants, concentrations of sodium borate and the addition of cyclodextrins (CD) to the buffer, and it was concluded that the less polar DM-β-CD added to 50 mM borate solution was the most suitable running buffer. In this condition, only aucubin and catalpol could not be separated, even with the addition of organic solvents and/or valine, urea and barium ion. The greatest advantage of capillary electrophoresis compared to HPLC analyses (the most commonly used technique) is its speed.

According to [132], capillary electrophoresis can be used to analyze a mixture of eleven iridoid glycosides: unedoside, harpagide, methyl catalpol, morroniside, asperuloside, griselinoside, catalpol, ketologanin, verbenalin, loganin and 10-cinnamoyl catalpol. Only loganin was found in the Apocynaceae family. For the analyses, iridoids were diluted in purified water. A Hewlett-Packard ^{3D}CE system coupled to a diode array detector and equipped with an air-cooling device was used. The fused-silica capillary tube measured 80 cm in length, 50 μm in I.D. and 375 μm in O.D. Distance to the detector was 71.5 cm only for UV detection (197 nm, 235 nm, 239 nm and 283 nm). When coupled to a mass spectrometer system (Bruker ESQUIRE) with an electrospray ionization source, the drying gas was nitrogen at 200 ºC and flow-rate 100 L/h. In this case, the distance between injector and UV detector was 20 cm. Other conditions: sample injection at 50 mbar for 5 s (only UV detection) or 25 s (with MS system); voltage, +20 kV; cartridge temperature, 25 °C; electrolyte solution, 20 mM ammonium acetate with 100 mM sodium dodecyl sulfate (SDS), pH 9.5; sheath liquid, 1 mM lithium acetate mixture to water/methanol (1:1 v/v) at a flow rate of 200 μL/h. When the MS system was used: scan range, 100-550 *m/z*; cut-off, 80 *m/z*; glass capillary exit, 95 V; skimmer, 32 V; electrospray voltage for the capillary, -4.0 kV; for the cylinder, -1,8 kV; for the end plate, -3.5 kV. In the comparison among the counterions sodium dodecyl sulfate (SDS), ammonium dodecyl sulfate and lithium dodecyl sulfate, diluted in water and running buffer, the best resolution for separating iridoid glucosides, lower noise in the MS system, and better repeatability and sensitivity were found with SDS in the running buffer. The volatility of ammonium acetate in buffer enables MS analyses, and concentrations higher than 20 mM did not represent better resolution. Quite the contrary, higher SDS concentrations furnished better results. In the study of the influence of pH, the best one was 9.5, although its influence in the range of 8.7-10.0 was lower than the SDS effect. Good linearity was observed for all the iridoids glucosides analyzed, but in different ranges.

The literature describes chromatographic techniques related to the characterization, isolation and purification of iridoids. Most reports show the open column technique as the principal technique used to isolate this class. Also, there have been few studies on counterflow and capillary electrophoresis chromatographies. In general, there has been little scientific investment in the area of obtaining iridoids of the Apocynaceae family, despite the great pharmacological importance of this class of constituents.

Author details

Ana Cláudia F. Amaral[1], Aline de S. Ramos[1], José Luiz P. Ferreira[1,2], Arith R. dos Santos[1], Deborah Q. Falcão[2], Bianca O. da Silva[4], Debora T. Ohana[1,5] and Jefferson Rocha de A. Silva[3]

*Address all correspondence to: acamaral@fiocruz.br

1 Laboratório de Plantas Medicinais e Derivados, Depto de Produtos Naturais, Farmanguinhos – FIOCRUZ, Manguinhos, Brazil

2 Faculdade de Farmácia – UFF, Niterói, Brazil

3 Laboratório de Cromatografia – Depto. de Química – UFAM, Japiim, Manaus, Brazil

4 Instituto de Pesquisas Biomédicas, Hospital Naval Marcílio Dias, Brazil

5 Fac. de Ciências Farmacêuticas – UFAM, Manaus, Brazil

References

[1] Bianco A, De Luca A, Mazzei RA, Nicoleit M, Passacantilli P, Limas RA. Iridoids of *Rauwolfia grandiflora*. Phytochemistry 1994; 35(6) 1485-1487.

[2] Dewick PM. Medical Natural Products: A biosynthetic Approach. 2º Edition. John Wiley & Sons, LTD; 2002.

[3] Sampaio-Santos MI, Kaplan MAC. Biosynthesis Significance of Iridoids in Chemosystematics. Journal of the Brazilian Chemical Society 2001; 12(2) 144-153.

[4] Oudin A, Courtois M, Rideau M, Clastre M. The Iridoid Pathway in *Catharanthus roseus* Alkaloid Biosynthesis. Phytochemical Reviews 2007; 6 259–276.

[5] Eisenreich W, Bacher A, Arigoni D, Rohdich F. Biosynthesis of Isoprenoids Via the Non-Mevalonate Pathway. Cellular and. Molecular. Life Sciences. 2004; 61 1401–1426.

[6] Hunter WN. The Non-Mevalonate Pathway of Isoprenoid Precursor Biosynthesis. Journal of Biological Chemistry 2007; 282 21573–21577.

[7] Contin A, Van der Heijden R, Lefeber AWM, Verpoorte R. The Iridoid Glucoside Secologanin is Derived from the Novel Triose Phosphate/Pyruvate Pathway in a *Catharanthus roseus* Cell Culture. FEBS Letters 1998; 434 413–416.

[8] Eichinger D, Bacher A, Zenk MH, Eisenreich, W. Analysis of metabolic pathways via quantitative prediction of isotope labeling patterns: a retrobiosynthetic 13C NMR study on the monoterpene loganin. Phytochemistry 1999; 51 223–236.

[9] Li H, Yang S, Wang H, Tian J, Gao W. Biosynthesis of the Iridoid Glucoside, Lamalbid, in *Lamium barbatum*. Phytochemistry 2010; 7 1690–1694.

[10] Inouye H, Uesato S. Biosynthesis of Iridoids and Secoiridoids. In: Progress in the Chemistry of Organic Natural Products, Vol. 50 (Herz W., Grisebach H., Kirby G. W., and Tamm Ch., eds.). Springer, New York, 1986 169-236.

[11] Jensen SR, Franzyk H, Wallander E. Review: Chemotaxonomy of the Oleaceae: Iridoids as Taxonomic Markers. Phytochemistry 2002; 60 213–231.

[12] Nagatoshi M, Terasaka K, Nagatsu A, Mizukami H. An Iridoid-Specific Glucosyltransferase from *Gardenia jasminoides.* JBC July, 2011: 1-19. Available from http://www.jbc.org/cgi/doi/10.1074/jbc.M111.242586 (accessed 20 September 2012).

[13] Galvez M, Martin-Cordero C, Ayuso MJ. Pharmacological Activities of Iridoids Biosynthesized by Route II. Studies in Natural Products Chemistry 2005; 32 365– 394.

[14] Cimanga K, Kambu K, Tona L, Hermans N, Apers S, Totte J, Pieters L, Vlietinek AJ. Antiamoebic Activity of Iridoids from *Morinda morindoides* Leaves. Planta Medica. 2006; 72 751-753.

[15] Yang XP, Li EW, Zhang Q, Yuan CS, Jia ZJ. Five New Iridoids from *Patrinia rupestris* Chemistry &. Biodiversity. 2006; 3 762-770.

[16] Tzakou O, Mylonas P, Vagias C, Petrakis PV. Iridoid Glucosides with Insecticidal Activity from *Galium melanantherum*. Zeitschrift fúr Naturforschung, C: Journal of Biosciences 2007; 62 597–602.

[17] da Silva VC, Giannini MJSM, Carbone V, Piacente S, Pizza C, Bolzani VS, Lopes MN. New Antifungal Terpenoid Glycosides from *Alibertia edulis* (Rubiaceae). Helvetica Chimica Acta 2008; 91 1355-1362.

[18] Tundis R, Loizzo MR, Menichini F, Statti GA, Menichini F. Biological and Pharmacological Activities of Iridoids: Recent Developments. Mini-Reviews in Medicinal Chemistry 2008; 8 399–420.

[19] Castillo L, Rossini C. Bignoniaceae Metabolites as Semiochemicals. Molecules 2010; 15 7090–7105.

[20] Zhang HJ, Rothwangl K, Mesecar AD, Sabahi A, Rong LJ, Fong HHS. Lamiridosins, Hepatitis C Virus Entry Inhibitors from *Lamium album*. Journal of Natural Products 2009; 72 2158–2162.

[21] Bai NS., HK, Ibarra A, Bily A, Roller M, Chen XZ, Rühl R. Iridoids from *Fraxinus excelsior* with Adipocyte Differentiation-Inhibitory and PPARα Activation Activity. Journal of Natural Products 2010 (73) 2–6.

[22] Dahlgren G. An updated angiosperm classification. Botanical Journal of the Linnean Society. 1989; 100: 197–203. doi: 10.1111/j.1095-8339.1989.tb01717.x (accessed 18 september 2012).

[23] Hegnauer, R. Chemotaxonomie der Pflanzen. Birkhäuser Verlag, Basel. 1986.

[24] Taskova R, Peev D, Handjieva N. Iridoid Glucosides of the Genus *Veronica* s.l. and their Systematic Significance. Plant Systematics and Evolution 2002; 231 1-17.

[25] Rivière C, Goossens L, Guerardel Y, Maes E, Garénaux E, Pommery J, Pommery N, Désiré O, Lemoine A, Telliez A, Delelis A, Hénichart J.P. Chemotaxonomic interest of iridoids isolated from a Malagasy species: *Perichlaena richardii.* Biochemical Systematics and Ecology 2011; 39: 797–825.

[26] Wink M, Waterman P. Chemotaxonomy in relation to molecular phylogeny of plants. In: Wink, M. (ed.), Biochemistry of plant secondary metabolism, Annual Plant Reviews. Boca Raton: Sheffield Academic Press and CRC Press. 1999: 300–341.

[27] Wink, M. Evolution of Secondary Metabolites from an Ecological and Molecular Phylogenetic Perspective. Phytochemistry 2003; 64 3–19.

[28] Dahlgren RMT. A revised System of Classification of Angiosperms. Botanical Journal of the Linnean Society. 1980; 80 91-124.

[29] Angiosperm Phylogeny Group III. An Update of the Angiosperm Phylogeny Group Classification for the Orders and Families of Flowering Plants: APG III. Botanical Journal of the Linnean Society 2009; 161 105–121.

[30] Taskova, RM, Gotfredsen CH, Jensen SR. Chemotaxonomy of Veroniceae and its Allies Plantaginaceae. Phytochemistry 2006; 67 286–301.

[31] Koch, I, Rapini A. Apocynaceae. In: Lista de espécies da flora do Brasil. Jardim Botânico do Rio de Janeiro, Rio de Janeiro. 2011. http:// floradobrasil.jbrj.gov.br/2011/ FB00004. (accessed 10 September 2012).

[32] Endress ME, Liede-Schumann S, Meve U. Advances in Apocynaceae: the Enlightment, an Introduction. Annals of the Missouri Botanical Garden 2007; 94 259–267.

[33] Rapini A. Taxonomy "Under Construction": Advances in the Systematic of Apocynaceae, with Emphasis on the Brazilian Asclepiadoideae. Rodriguésia 2012; 63(1): 75-88.

[34] Meve U. Species Numbers and Progress in Asclepiad Taxonomy. Kew Bulletin 2002; 57 459–464.

[35] The International Plant Names Index. http://www.inpi.org/ (accessed 04 October 2012).

[36] Tropicos. org. Missouri Botanical Garden. http://www.tropicos.org (accessed 04 October 2012).

[37] Siddiqui BS, Naeed A, Begum S, Siddiqui S. Minor Iridoids from the Leaves of *Plumeria obtusa*. Phytochemistry 1994; 37(3) 769-771.

[38] Begun S, Naeed A, Siddiqui BS, Siddiqui S. Chemical Constituents of the Genus *Plumeria*. Journal of the Chemical Society of Pakistan 1994; 16(4) 280-299.

[39] Abe F, Möri T, Yamauchi T. Iridoids of Apocinaceae. III. Minor Iridoids from *Allamanda neriifolia*. Chemical Pharmaceutical Bulletin 1984; 32(8) 2947-2956.

[40] Boros CA, Stermitz FR. Iridoids. An Update Review, Part II. Journal of Natural Products 1991; 54(5) 1173-1246.

[41] Kupchan SM, Dessertine AL, Blaylock BT, Bryan RF. Isolation and Structural Elucidation of Allamandin, an Antileukemic Iridoid Lactone from *Allamanda cathartica*. Journal of Organic Chemistry 1974; 39(17) 2477-2482.

[42] El-Naggar LJ, Beal JL. Iridoids. A Review. Journal of Natural Products 1980; 43(6) 649-707.

[43] Sharma G, Chahar MK, Dobhal S, Sharma N, Sharma TC, Sharma MC, Joshi YC, Dobhal MP. Phytochemical Constituents, Traditional Uses and Pharmacological Properties of the Genus *Plumeria*. Chemical & Biodiversity 2011; 8(8) 1357-1369.

[44] Kardono LBS, Tsauri S, Padmawinata K, Pezzuto JM, Kinghorn AD. Cytotoxic Constituents of the Bark of *Plumeria rubra* Collected in Indonesia. Journal of Natural Products 1990; 53(6) 1447-1455.

[45] Vanderlei MF, Silva MS, Gottlieb HF, Braz-Filho R. Iridoids and Triterpenes from Himatanthus phagedaenica: the Complete Assignment of the Proton and Carbon-13 NMR Spectra of Two Iridoid Glycosides. Journal of the Brazilian Chemical Society 1991; 2(2) 51-55.

[46] Al-Hazimi HMG, Alkhathlan H. Naturally Occurring Iridoids During the Period 1990-1993. Journal of the Chemical Society of Pakistan 1996; 18(4) 336-357.

[47] Anderson JE, Chang CJ, McLaughlin JL. Bioactive Components of *Allamanda schottii*. Journal of Natural Products 1988; 51(2) 307-308.

[48] Kuigoua GM, Kouam SF, Ngadjui BT, Schulz B, Green IR, Choudhary MI, Krohn K. Minor Secondary Metabolic Products from the Stem Bark of *Plumeria rubra* Linn. Displaying Antimicrobial Activities. Planta Medica 2010; 76(6) 620-625.

[49] Lopes RK, Ritter MR, Rates SMK. Revisão das Atividades Biológicas e Toxicidade das Plantas Ornamentais mais Utilizadas no Rio Grande do Sul, Brasil. Brazilian Journal of Biosciences 2009; 7(3) 305-315.

[50] Morel AF, Graebner IB, Porto C, Dalcol II. Study on the Antimicrobial Activity of *Himatanthus sucuuba*. Fitoterapia 2006; 77(1) 50-53.

[51] Amaral ACF, Ferreira JLP, Pinheiro MLB, Silva JR de A. Monograph of *Himatanthus sucuuba*, a Plant of Amazonian Folk Medicine. Pharmacognosy Reviews 2007; 1(2) 305-313.

[52] Yu YL, Li X, Ke CA, Tang CP, Yang XZ, Li XQ, Ye Y. Iridoid Glucosides from *Allamanda neriifolia*. Chinese Chemical Letters 2010; 21(6) 709-711.

[53] Dinda B, Debnath S, Banik R. Naturally Occurring Iridoids and Secoiridoids. An Update review, Part 4. Chemical Pharmaceutical Bulletin 2011; 59(7) 803-833.

[54] Shen YC, Chem CH. Allaneroside and Bioactive Iridoid Glycosides from *Allamanda neriifolia*. Taiwan Yaoxue Zazhi 1986; 38(4) 203-213.

[55] Abe F, Yamauchi T, Wan ASC. Studies on *Cerbera*: VIII. Normonoterpenoids and their Allopyranosides from the Leaves of *Cerbera* Species. Chemical & Pharmaceutical Bulletin 1989; 37(10) 2639-2642.

[56] Shen LR, Jin SM, Yin BW, Du XF, Wang YL, Huo CH. Chemical Constituents of Plants from the Genus *Cerbera*. Chemistry & Biodiversity 2007; 4(7) 1438-1449.

[57] Thomas PS, Kanaujia A, Ghosh D, Duggar R, Katiyar CK. Alstonoside, a Secoiridoid Glucoside from *Alstonia scholaris*. Indian Journal of Chemistry, section B: Organic Chemistry Including Medicinal Chemistry 2008; 47B(8), 1298-1302.

[58] Dinda B, Chowdhury DR, Mohanta BC. Naturally Occurring Iridoids, Secoiridoids and Their Bioactivity. An Update Review, Part 3. Chemical Pharmaceutical Bulletin 2009; 57(8) 765-796.

[59] Topcu G, Che CT, Cordell GA, Ruangrungsi N. Iridolactones from *Alyxia reinwardtii*. Phytochemistry 1990; 29(10) 3197-3199.

[60] Feng T, Cai XH, Du ZZ, Luo XD. Iridoids from the Bark of *Alstonia scholaris*. Helvetica Chimica Acta 2008; 91(12) 2247-2251.

[61] Yamauchi T, Abe F, Wan ASC. 10-*O*-Benzoyltheveside and 10-Dehydrogeniposide from the Leaves of *Cerbera manghas*. Phytochemistry 1990; 29(7) 2327-2328.

[62] Abe F, Yamauchi T. 10-Carboxyloganin, Normonoterpenoid Glucosides and Dinormonoterpenoid Glucosides from the Leaves of *Cerbera manghas* (Studies on *Cerbera*. X). Chemical & Pharmaceutical Bulletin 1996; 44(10) 1797-1800.

[63] Marini-Bettolo GB, Nicoletti M, Messana I, Patamia M, Galeffi C, Oguakwa JU, Portalone G, Vaciago A. Research on African Medicinal Plants - IV. Boonein, a New C-9 Terpenoid Lactone from *Alstonia boonei*: A Possible Precursor in the Indole Alkaloid Biogenesis. Tetrahedron 1983; 39(2) 323-329.

[64] Adotey JPK, Adukpo YDB, Armah FA. A Review of the Ethnobotany and Pharmacological Importance of *Alstonia boonei* De Wild (Apocynaceae). ISRN Pharmacology 2012; Article ID 587160, 9 pages. http://www.hindawi.com/isrn/pharmacology/2012/587160/ (accessed 15 september 2012). doi: 10.5402/2012/587160.

[65] Patil CD, Patil SV, Borase HP, Salunke BK, Salunkhe RB. Larvicidal Activity of Silver Nanoparticles synthesized Using *Plumeria rubra* Plant Latex Against *Aedes aegypti* and *Anopheles stephensi*. Parasitological Research 2012; 110(5) 1815-1822.

[66] Abe F, Chen RF, Yamauchi T. Studies on the Constituents of *Plumeria*. Part I. Minor Iridoids from the Roots of *Plumeria acutifolia*. Chemical & Pharmaceutical Bulletin 1988; 36(8) 2784-2789.

[67] Abe F, Okabe H, Yamauchi T. Studies on *Cerbera*: II. Cerbinal and its Derivatives, Yellow Pigments in the Bark of *Cerbera manghas* L.. Chemical & Pharmaceutical Bulletin 1977; 25(12) 3422-3424.

[68] Zhang X, Zhang J, Liu M, Pei Y. Chemical Components in Cortex of *Cerbera manghas* L. Zhongcaoyao 2008; 39(8) 1138-1140.

[69] Ali FI, Hashmi IL, Siddiqui BS. A novel iridoid from *Plumeria obtusa*. Natural Product Communications 2008; 3(2) 125-128.

[70] Barreto A de S, Amaral ACF, Silva JR de A, Schripsema J, Rezende CM, Pinto AC. Ácido 15-desmetilisoplumierídeo, um Novo Iridóide Isolado das Cascas de *Plumeria rubra* e do Látex de *Himatanthus sucuuba*. Química Nova 2007; 30(5) 1133-1135.

[71] Hassan EM, Shahat AA, Ibrahim NA, Vlietinck A, Apers S, Pieters L. A New Monoterpene Alkaloid and other Constituents of *Plumeria acutifolia*. Planta Medica 2008; 74(14) 1749-1750.

[72] Saleem M, Akhtar N, Riaz N, Ali MS, Jabbar A. Isolation and Characterization of Secondary Metabolites from *Plumeria obtusa*. Journal of Asian Natural Products Research 2011; 13(12) 1122-1127.

[73] Silva JR de A, Amaral ACF, Silveira CV, Rezende CM, Pinto AC. Quantitative Determination by HPLC of Iridoids in the Bark and Latex of *Himatanthus sucuuba*. Acta Amazonica 2007; 37(1) 119-122.

[74] Albers-Schoenberg G, Schmid H. Über Die Struktur von Plumericin, Isoplumericin, β-Dihydroplumericinic Acid. Helvetica Chimica Acta 1961; 44(181) 1447-1473.

[75] Elsässer B, Krohn K, Akhtar MN, Florke U, Kouam SF, Kuigoua MG, Ngadjui BT, Abegaz BM, Antus S, Kurtán T. Revision of the Absolute Configuration of Plumericin and Isoplumericin from *Plumeria rubra*. Chemistry & Biodiversity 2005; 2(6) 799-808.

[76] Vankata Rao E, Anjaneyulu TSR. Chemical Components of the Bark of *Plumeria rubra*. Indian Journal of Pharmacy 1967; 29(9) 273-274.

[77] Tandon SP, Tiwarik P, Rathore YKS. Chemical Constituents of the Decorticated Stem of *Plumeria rubra* Linn. Proceedings of the National Academy of Sciences, India Section A 1976; 46(2) 109-110.

[78] Grumbach A, Schmid H, Bencze W. An Antibiotic from *Plumeria acutifolia*. Experientia 1952; 8 224-225.

[79] Schmid H, Bencze W. Über die Konstitution des Fulvoplumierins I. Helvetica Chimica Acta 1953; 36(27) 205-214.

[80] Rangaswami S, Venkata Rao E, Suryanarayana M. Chemical Examination of *Plumeria acutifolia*. Indian Journal of Pharmacy 1961; 23 122-124.

[81] Wanner H, Zorn-Ahrens V. Distribution of Plumieride in *Plumeria acutifolia* and *Plumeria bracteata*. Berichte de Schweizerischen Botanischen Gesellschaft 1972; 81 27-39.

[82] Mahran GH, Abdel-Wahab, SM, Ahmed MS Detection and Isolation of the Antibiotic Principles of *Plumeria rubra* L. and *Plumeria rubra* L. var. *alba*. Growing in Egypt. Bulletin of the Faculty of Pharmacy (Cairo University) 1975; 12(2) 151-60.

[83] Albers-Schönberg, G, Philipsborn WV, Jackman LM, Schmid H. Die Struktur de Fulvoplumierins. Helvetica Chimica Acta 1962; XLV(164) 1406-1408.

[84] Coppen JJW, Cobb AL. The Occurrence of Iridoids in *Plumeria* and *Allamanda*. Phytochemistry 1983; 22(1) 125-128.

[85] Hamburger MO, Cordell GA, Ruangrungsi N. Traditional Medicinal Plants of Thailand XVII. Biologically Active Constituents of *Plumeria rubra*. Journal of Ethnopharmacology 1991; 33(3) 289-292.

[86] Sharma U, Singh D, Kumar P, Dobhal MP, Singh S. Antiparasitic Activity of Plumericin & Isoplumericin Isolated from *Plumeria bicolor* against *Leishmania donovani*. Indian Journal of Medical Research 2011; 134(5) 709-716.

[87] Barreto A de S, Carvalho MG, Neri I, Gonzaga L, Kaplan MAC. Chemical Constituents from *Himatanthus articulata*. Journal of the Brazilian Chemical Society 1998; 9(5) 430-434.

[88] Castillo D, Arevalo J, Herrera F, Ruiz C, Rojas R, Rengifo E, Vaisberg A, Lock D, Lemesre JL, Gornitzka H, Sauvain M. Spirolactone Iridoids might be responsible for the Antileishmanial Activity of a Peruvian Traditional Remedy Made with *Himatanthus sucuuba* (Apocynaceae). Journal of Ethnopharmacology 2007; 112(2) 410-414.

[89] Abdel-Kader MS, Wisse J, Evans R, van der Werff H, Kingston DGI. Bioactive Iridoids and a New Lignan from *Allamanda cathartica* and *Himatanthus fallax* from the Suriname Rainforest. Journal of Natural Products 1997; 60(12) 1294-1297.

[90] Silva JR de A, Rezende CM, Pinto AC, Pinheiro MLB, Cordeiro MC, Tamborini E, Young CM, Bolzani V da S. Ésteres Triterpênicos de *Himatanthus sucuuba* (Spruce) Woodson. Química Nova 1998; 21(6) 702-704.

[91] Miranda ALP, Silva JR de A, Rezende CM, Neves JS, Parrini SC, Pinheiro MLB, Cordeiro MC, Tamborini E, Pinto AC. Anti-inflammatory and Analgesic Activities of the Latex Containing Triterpenes from *Himatanthus sucuuba*. Planta Medica 2000; 66(3) 284-286.

[92] Silva JR de A, Rezende CM, Pinto AC, Amaral ACF. Cytotoxicity and Antibacterial Studies of Iridoids and Phenolic Compounds Isolated from the Latex of *Himatanthus sucuuba*. African Journal of Biotechnology 2010; 9(43) 7357-7360.

[93] Vilegas JHY, Hachich EM; Garcia M, Brasileiro A, Carneiro MAG, Campos VLB. Antifungal Compounds from Apocynaceae Species. Revista Latinoamericana de Química 1995; 23(2) 73-75.

[94] Vilegas JHY, Hachich EM, Garcia M, Brasileiro A, Carneiro MAG, Campos VLB. Antifungal compounds from Apocynaceae species. Revista Latinoamaricana de Química 1992; 23(1) 44-45.

[95] Ferreira JLP, Amaral ACF, Araújo RB, Carvalho JR, Proença CEB, Fraga SAPM, Silva JR de A. Pharmacognostical Comparison of three Species of *Himatanthus*. International Journal of Botany 2009; 5(2) 171-175.

[96] Bobbitt J.M., Segebarth K.P. The Iridoid Glycosides and Similar Substances. In: Battersby A.R. and Taylor W.I. (eds.) Cyclopentanoid Terpene Derivatives. New York: Marcel Dekker; 1969. p25-31.

[97] Ye G, Li ZX, Xia GX, Peng H, Sun ZL, Huang CG. A New Iridoid Alkaloid from the Flowers of *Plumeria rubra* L. cv. *acutifolia*. Helvetica Chimica Acta 2009; 92(12) 2790-2794.

[98] Stephens PJ, Pan JJ, Krohn K. Determination of the Absolute Configurations of Pharmacological Natural Products via Density Functional Theory Calculations of Vibrational Circular Dichroism: The New Cytotoxic Iridoid Prismatomerin. Journal of Organic Chemistry 2007; 72(20) 7641-7649.

[99] Little JE, Johnstone DB. Plumericin: An Antimicrobial Agent from *Plumeria multiflora*. Archives of Biochemistry 1951; 30 445-452.

[100] Wood CA, Lee K, Vaisberg AJ, Kingston DGI, Neto CC, Hammond GB. A Bioactive Spirolactone Iridoid and Triterpenoids from *Himatanthus sucuuba*. Chemical & Pharmaceutical Bulletin 2001; 49(11) 1477-1478.

[101] Dobhal MP, Li G, Gryshuk A, Bhatanager AK, Khaia SD, Joshi YC, Sharma MC, Oseroff A, Pandey RK. Structural Modifications of Plumieride Isolated from *Plumeria bicolor* and the Effect of these Modifications on In Vitro Anticancer Activity. Journal of Organic Chemistry 2004; 69(19) 6165-6172.

[102] Sousa LMA, Neto, RLM, Schmidt FN, Oliveira MR. Anti-mitotic activity Towards Sea Urchin Eggs of Dichloromethane Fraction Obtained from *Allamanda schottii* Pohl (Apocynaceae). Brazilian Journal of Pharmacognosy 2009; 19(2A) 349-352.

[103] Waltenberger B, Rollinger JM, Griesser UJ, Stuppner H, Gelbrich T. Plumeridoid C from the Amazonian Traditional Medicinal Plant *Himatanthus sucuuba*. Acta Crystallgraphica Section C: Crystal Structure Communications 2011; 67(10) 409-412.

[104] Mahran GH, Abdel Wahab SM, Salah AM. Isolation and Quantitative Estimation of Plumieride from the different Organs *Plumeria rubra* and *Plumeria rubra* var. *alba*. Planta Medica 1974; 25(3) 226-230.

[105] Mahran GH, Abdel-Wahab SM, Ahmed MS. Isolation and Estimation of Plumieride from Different Organs of *Plumeria rubra* L. and *Plumeria rubra* L. var. *alba.* Bulletin of the Faculty of Pharmacy (Cairo University) 1975a; 12(2)133-139.

[106] Schmid IH, Bickel H, Meijer ThM. Zur Kenntnis des Plumierids. 1. Mitteilung. Helvetica Chimica Acta 1952; 35(1)415-427.

[107] Yeowell DA, Schmid H. Zur Biosynthese des Plumierids. Experientia 1964; 20(5) 250-252.

[108] Rangaswami S, Venkata Rao E. Chemical Components of *Plumeria alba*. Proceedings – Indian Academy of Sciences Section A 1960; 52A 173-181.

[109] Afifi MS, Salama OM, Gohar AA, Marzouk NA. Iridoids with Antimicrobial Activity from *Plumeria alba*. Bulletin of Pharmaceutical Sciences, Assiut University 2006; 29(1) 215-223.

[110] Adam G, Khoi NH, Bergner C, Lien NT. Natural Products from Vietnamese Plants. Part. 3. Plant Growth Inhibiting Properties of Plumieride from *Plumeria obtusifolia*. Phytochemistry 1979; 18(8) 1399-1400.

[111] Halpern O, Schmid H. Zurkenntnis de Plumierids. 2. Mitteilung. Helvetica Chimica Acta 1958; 41(4) 1109-1154.

[112] Inouye H, Uobe K, Hirai M, Masada Y, Hashimoto K. Studies on Monoterpene Glucosides and Related Natural Products. XXX. Gas Chromatography and Gas Chromatography-Mass Spectrometry of Iridoid and Secoiridoid Glucosides. Journal of Chromatography 1976; 118(2) 201-216.

[113] Grignon-Dubois M, Rezzonico B, Usubillaga A, Vojas LB. Isolation of Plumieride from *Plumeria inodora*. Chemistry of Natural Compounds 2005; 41(6) 730-731.

[114] Gupta RS, Bhatnager AK, Joshi YC, Sharma R, Sharma A. Effects of plumieride, an iridoid on spermatogenesis in male albino rats. Phytomedicine 2004; 11(2-3) 169-174.

[115] Rebouças S de O, Grivicich I, dos Santos MS, Rodriguez P, Gomes MD, de Oliveira SQ, da Silva J, Ferraz ABF. Antiproliferative Effect of a Traditional Remedy, *Himatanthus articulatus* Bark, on Human Cancer Cell Lines. Journal of Ethnopharmacology 2011; 137(1) 926-929.

[116] Coppen JJW. Iridoids with Algicidal Properties from *Allamanda cathartica*. Phytochemistry 1983; 22(1) 179-182.

[117] Yamauchi T, Abe F, Taki M. Protoplumericin, an Iridoid Bis-Glucoside in *Allamanda neriifolia*. Chemical and Pharmaceutical Bulletin 1981; 29(10) 3051-3055.

[118] Boonclarm D, Sornwatana T, Arthan D, Kongsaeree P, Svasti J. β-Glucosidase Catalyzing Specific Hydrolysis of an Iridoid β-Glucoside from *Plumeria obtuse*. Acta Biochimica et Biophysica Sinica 2006; 38(8) 563-570.

[119] Siddiqui BS, Ilyas F, Rasheed M, Begum S. Chemical Constituents of Leaves and Stem Bark of *Plumeria obtusa*. Phytochemistry 2004; 65(14) 2077-2084.

[120] Ye G, Yang YL, Xia GX, Fan MS, Huang CG. Complete NMR Spectral Assignments of Two New Iridoid Diastereoisomers from the Flowers of *Plumeria rubra* L.CV. *acutifolia*. Magnetic Resonance in Chemistry 2008; 46(12) 1195-1197.

[121] Kitagawa I, Shibuya H; Baek IN, Yokokawa Y, Nitta A, Wiriadinata H, Yoshikawa M. Pulosarioside, a new bitter trimeric-iridoid diglucoside, from an Indonesian jamu, the bark of Alyxia reinwardtii BL. (Apocynaceae). Chemical & Pharmaceutical Bulletin 1988; 36(10) 4232-4235.

[122] Sticher O. Theveside, a New Iridoid Glucoside from *Thevetia peruviana* (Apocynaceae). Tetrahedron Letters 1970; 11(36) 3195-3196.

[123] Sticher O. Theveside, a New Iridoid-Glucoside from *Thevetia peruviana* (*Thevetia neriifolia*) (Apocynaceae). Pharmaceutica Acta Helvetiae 1971; 46(3) 156-166.

[124] Inouye H, Nishimura T. Iridoids Glucosides of *Cerbera manghas*. Phytochemistry 1972; 11(5) 1852.

[125] Abe F., Chen R.F., Yamauchi T, Ohashi H. Iridoids from the Roots of *Thevetia peruviana*. Chemical & Pharmaceutical Bulletin 1995b; 43(3) 499-500.

[126] Sticher O, Schmid H. Theviridosid, ein iridoid-glucosid aus *Thevetia peruviana* (Pers.) K. Schum. (*Thevetia neriifolia* Juss.) Vorläufige Mitteilung. Helvetica Chimica Acta 1969; 52(2) 478-481.

[127] Abe F, Yamauchi T, Xahara S, Nohara T. Minor Iridoids from *Thevetia peruviana*. Phytochemistry 1995; 38(3)793-794.

[128] Changwichit K, Khorana N, Suwanborirux K, Waranuch N, Limpeanchob N, Wisuitiprot W, Suphrom N, Ingkaninan K. Bisindole alkaloids and secoiridoids from *Alstonia macrophylla* Wall. ex G. Don. Fitoterapia 2011; 82(6) 798-804.

[129] Zhu W-M, He H-P, Fan L-M, Shen Y-M, Zhou J, Hao X-J. Components of Stem barks of *Winchia calophylla* A. DC. and their bronchodilator activities. Acta Botanica Sinica 2005; 47(7) 892-896.

[130] Basu D, Chaterjee A. Ocurrence of Plumericin in *Nerium indicum*. Indian Journal of Chemistry 1973; 11 297.

[131] Wu H-K, Chuang W-C, Sheu S-J. Separation of nine iridoids by capillary electrophoresis and high-performance liquid chromatography. Journal of Chromatography A 1998; 803(1-2) 179-187.

[132] Suomi J, Wiedmer SK, Jussila M, Riekkola M-L. Analysis of eleven iridoid glycosides by micellar electrokinetic capillary chromatography (MECC) and screening of plant samples by partial filling (MECC)-electrospray ionization mass spectrometry. Journal of Chromatography A 2002; 970(1-2) 287-296.

Chapter 7

Analysis of the Presence of the Betulinic Acid in the Leaves of *Eugenia florida* by Using the Technique GC/MS, GC/FID and HPLC/DAD: A Seasonal and Quantitative Study

Alaíde S. Barreto, Gláucio D. Feliciano,
Cláudia Cristina Hastenreiter da Costa Nascimento,
Carolina S. Luna, Bruno da Motta Lessa,
Carine F. da Silveira, Leandro da S. Barbosa,
Ana C. F. Amaral and Antônio C. Siani

Additional information is available at the end of the chapter

http://dx.doi.org/10.5772/55868

1. Introduction

The betulinic acid (Figure 1) is a known triterpenoid isolated from various organs and species of plants, including flowering *Eugenia* DC [Junges, 1999]. This metabolite shows inhibitory activity on growth of human melanoma cells [Pisha *et al.*, 1995], and replication of the AIDS virus [Evers et al. 1996; Soler *et al.*, 1996]. In additional betulinic acid derivatives [Chatterjee *et al.*, 2000; Galgon *et al.*, 2005] induced cell apoptosis of human melanoma. This specificity in melanoma cells makes the substance compared to complex molecules such as taxol, the most promising anticancer drug [Pisha *et al.* 1995]. However, their action is limited only neuroblastomas and melanoma cells and is not active against other cancer cells [Chatterjee, 2000, Pezzuto *et al.*, 1999; Pisha, 1995; Mayauxet *al.*, 1994]. The betulinic acid also has antibacterial property and inhibits the growth of colonies of *Escherichia coli* and *Staphylococcus aureus*.

Despite all of betulinic acid pharmacological potential, it is obtained by extraction of barks or core of some plant species or by synthetic processes, e.g. using the betulin (alcohol triterpene) as a synthetic intermediate isolated from the bark of *Betula alba* and *Betula pendula* [Galgon., *et al.* 1999]. Therefore, research is necessary to identify new natural sources, which produce large quantities of substance easily renewable parts of the plant (leaf) thereby not affecting plant growth, development of chromatographic methods rapid and easy manipulation studies to identify the seasonal best months of collection.

As part of a program conducted in our laboratory involving search for new sources of bioactive metabolites from Brazilian plants, we investigated the leaves of *Eugenia florida*. This species belongs to the family Myrtaceae. Compounds such as flavonoids, triterpenes, tannins and especially essential oils constituted of monoterpenes and sesquiterpenes have already been isolated from the genus Eugenia [Lunardi, et al., 2001]. The species of this family are widely distributed in the Brazilian forests, much of it is popularly known for its edible fruits, wood, essential oils or ornamental purposes [Consolini *et al.*, 1999, Costa *et al.*, 2005; Siani *et al.*, 2000]. The most important genera of this family are: *Melaleuca, Eucalyptus, Psidium* and *Eugenia* [Siani *et al.*, 2000].

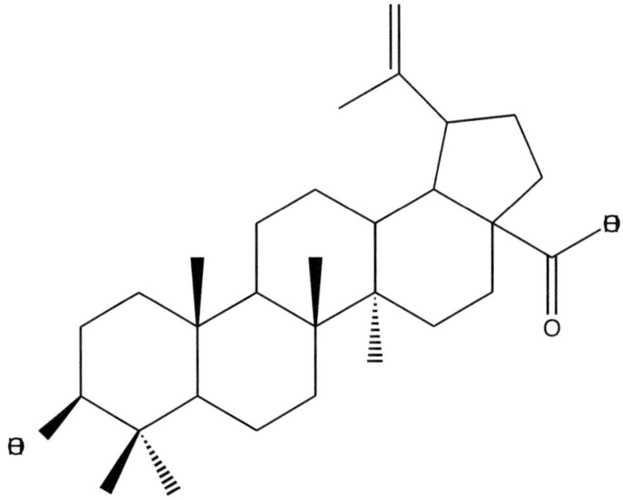

Figure 1. Betulinic acid

1.1. Seasonal variation

Since the fourth century B.C. there are reports of procedures for the collection of medicinal plants. The executioners Greeks, e.g., they collected their samples of poison hemlock (*Conium maculatum*) morning when levels are higher alkaloid coniina [Robinson, 1974]. Temporal and spatial variations in the total content, as well as the relative proportions of secondary metabolites in plants occur at different levels and, despite the existence of a genetic control, the expression may undergo changes resulting from the interaction of biochemical processes, physiological, ecological and evolutionary [Gobbo, Lopes, 2007]. In fact, the secondary metabolites represent a chemical interface between plants and the surrounding environment. Therefore, their synthesis is often affected by environmental conditions [Gobbo, Lopes, 2007].

Several factors that can coordinate or alter the rate of production of secondary metabolites, genetic factors, physical environment, collection method (date, time, etc.), drying conditions and transport, storage, pH of the soil, growing conditions, nutrient soil, plant part used, interactions between plants, the presence of microorganisms, can directly affect the concentration of the chemical components of each species [Silva, 1996]. Some factors have correlations with each other and not act alone, and may jointly affect the secondary metabolism, e.g. development and seasonality, rainfall and seasonality, temperature and altitude, among others [Gobbo, Lopes, 2007]. It should also be noted that, often, the changes may result from leaf development and / or appearance of new organs concomitant with constancy in the total content of secondary metabolites. This may cause decrease in the concentration of these metabolites by dilution may, however, result in a higher total amount due to the increase of biomass. [Hendriks *et al.*, 1997; Spring, Bienert, Klemt 1987]

The seasonal variation (collection of plants at different times or seasons), is a very significant factor in the percentage and production of secondary metabolites [Mitscher, Pillai, Shankel, 2000; Navarro *et al.*, 2002]. Due to its importance, it is necessary a study of seasonality, when working with medicinal plants [Mitscher, Pillai, Shankel, 2000; Navarro *et al.*, 2002].

This work had as main objectives to develop a protocol for the quantification of betulinic acid present in the leaves of *Eugenia florida* by using the technique GC/MS, GC/FID and HPLC/ DAD and through studies demonstrates the potential of this seasonal vegetable production as a source of natural metabolite.

2. Material and methods

2.1. General experimental procedures

1H and 13C NMR spectra were recorded on a Brucker AM-200 and 500MHz) chemical shifts are given in d values referred to internal tetramethlysilane (TMS), EIMS (MS Agilent 5973; 70eV) and Infrared (IV) spectra were recorded on a Nicolet spectrophotometer with Fourier transform Model Magna–IR 760 wavelengths are expressed in reciprocal centimeter (cm^{-1}).

HPLC analysis were performed at room temperature using the following system: second pump system (Shimadzu LC10AD model, Japan) a photodiode array detector (Shimadzu, SPD10ADVP model), an auto injector (Shimadzu SIL10ADVP model) an oven (Shimadzu, CTO 8-A model) and under the following conditions: Shimpack C-8 (1cm x 4.6mm i.d.) guard column and C-18 column (25cmx4.6mm i.d.; 5μm particle size) was from Zorba Zx provided by Agilent Technologies (USA). The system was controlled by Class Vp (Shimadzu, Japan) software 5.16. The gradient mobile phase was carried acetonitrile (Tedia, Brazil) and water HPLC grade. The mobile phase was degassed with helium and the flow rate adjusted to 1mL. min^{-1}. The water was acidified with TFA (0.05% v/v). All samples were injected automatically (10μ L) in triplicate.

In GC/FID experiments, 1mg each extracts [August 2009 to July 2010] and standard (Carl Roth – Karlsruhe, Germany) were transferred into glass vial and submitted methlylation with

CH_2N_2 with 100% yield. Those samples were dissolved in CH_2Cl_2 in concentrations 1.4mg/mL and injected for GC-FID and GC-MS analysis. The identification of methylated betulinic acid in extracts was done with use Wiley and NBS peak matching library search system. Authentic standard of the betulinic acid and data reported in the literature were also used for further identification as described.

2.2. Plant material

Healthy leaves of *Eugenia florida* and adults were collected during 12 months [August 2009 to July 2010] on the campus Oswaldo Cruz Foundation, state of Rio de Janeiro. The specie identification was carried out by biologist Sergio Monteiro of Oswaldo Cruz Foudantion [Laboratory of Production and Processing of Raw Plant (LPBMPV)], and a voucher specimen was deposited in the Herbarium of the Botanical Garden of Rio de Janeiro with the number RB 328061.

2.3. Methylation

A solution of diazomethane (CH_2N_2) in ether was prepared and added (excess) in drops of the solutions of extracts, EF-1 and standard (1mg) $CHCl_3$ or MeOH. The resulting solutions were allowed to stand for 12 hours and the ether removed by passing a stream of N_2 [Leonard, Lygo, Procter, 1995].

2.4. Betulinic acid quantification

GC analysis was performed on 6890N (Agilents Technologies, Network series) equipped whit a HP-5 column (30 x 0,25mm; 0,25μm liquid phase). Oven temperature program of 70°C - 300°C at 5°C/min; carrier gas: helium 11,3L/min; split mode of (20:1) and finally held for 30min. The mass spectrometer unit was performed with the same conditions the GC analysis. The calibration curve of the GC / FID was made in triplicate from different concentrations of esterified betulinic acid standard (0.1 to 1μg.mL^{-1}) and the curve was constructed using the average values of the detector response. The detector response was linear to the concentration internal of 0.1 to 1μg.mL^{-1} ($r^2 = 0.999$, Figure 2a).

HPLC grade acetonitrile was purchased from TEDIA (Brazil); 0,05% TFA (Trifluoroacetic acid) from Vetec (Brazil). Water was purified by Milli-Q_{plus} system from Milipore (Milford, MA, USA). Betulinic acid was purchased from Carl Roth (Karlsruhe, Germany with 99%). The 30mg ethanol extracts were then dissolved in 5mL of mobile phase. The mobile phase consisted of a gradient of 0,05% aqueous trifluoroacetic acid: acetonitrile delivered at a 1.0mL.min^{-1} as follows initial (t= 0 min) 30:70, linear gradient over 20min to 15:85, linear gradient over 10min to 100:0 and a new linear gradient over 20min (30:70); 40min as total time of analysis. Flow rate was 1mL.min^{-1}. Quantification was performed using the detector set at a wavelength of 210nm. Injection volume was 30μl. The peak of betulinic acid was identified in each chromatogram from of the ethanol extracts monthly (twelve months) with the help of injection of the standard solution of betulinic acid or comparison of the UV spectrum. The calibration curve of the HPLV-UV was made in triplicate from different concentrations of betulinic acid standard (0.1 to 1μg.mL^{-1}) and the curve was constructed using the average values of the detector response. The detector response was linear to the concentration internal of 0.1 to 0,5μg.mL^{-1} (r^2 = 0.9994, Figure 2b).

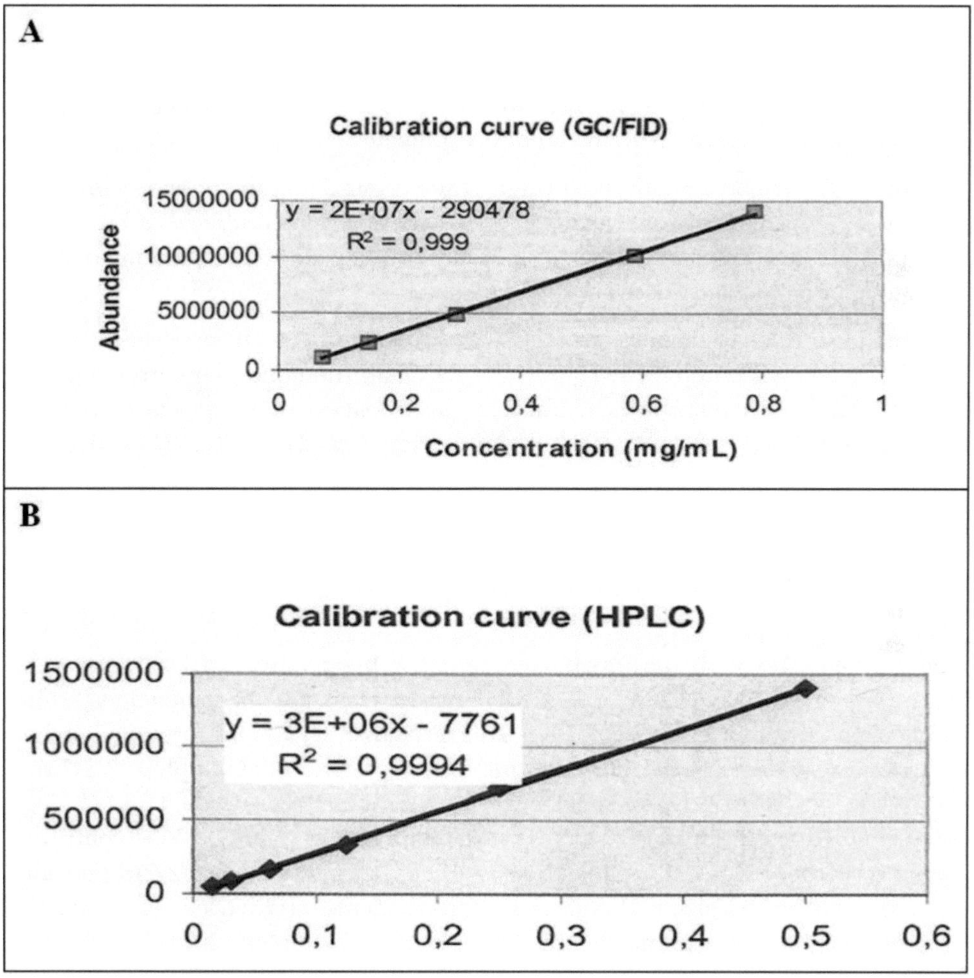

Figure 2. Calibration curve of Betulinic acid, A: GC-FID and B: HPLC-UV

3. Results and discussion

3.1. Extraction of betulinic acid

The leaves of *Eugenia florida* (17.1 kg) were dried at 400C, ground and subjected to soxhlet extraction with ethanol. The diluted extract was removed under reduced pressure (4.7 g). An aliquot of the methanol extract (200 mg) was dissolved in methanol (20ml) and recrystallized using mixtures of $CHCl_3$ and MeOH. Recrystallization was obtained a white crystal (EF – 1; 50mg).

EF-1 was analyzed by spectrophotometer 1H and 13C NMR (Bruker AC 200, 200MHz) using as solvent chloroform ($CDCl_3$) and methanol (CD_3OD) deuterated at a ratio of 9:1 to tetramethylsilane (TMS) as internal reference standard. An aliquot of EF-1 (5 mg) was methylated with diazomethane and subjected to mass spectrometry (MS; Agilent Technologies). The spectral data obtained were compared with the literature [Oliveira *et al.*, 2006].

The substance showed an EF-1 in the form of white crystals and the IR spectrum showed a broad band at $3450cm^{-1}$ by a characteristic of hydroxyl groups and acid, a broad band at $2942cm^{-1}$ one of alkyl groups and bands at $1686cm^{-1}$ and $1639cm^{-1}$ corresponding respectively to the axial deformation of carbonyl acid and alkene.

The information that led to elucidation of the structure was obtained from experiments nuclear magnetic resonance spectra [DEPT, HMQC, ^1H-^1H COSY (homonuclear correlation spectroscopy) and HMBC experiment] which indicate a known pattern of the terpenes series lupanos (Nick *et al.*, 1994, Mahato, Kundu 1994; Budzikiewicz *et al.*, 1964). The ^1H NMR spectrum showed two signals of multiplet in δ_H 4.69 and 4.58, referring to vinyl hydrogen (H-20), δ_H1.66 a signal corresponding to the methyl group bonded to carbon and fifth signals sp^2 corresponding to the methyl tertiary (δ_H 0.74; 0.85, 0.94, 0.96 and 1.00). The ^{13}C NMR spectrum confirmed the presence of signals in vinyl 152.02 and 110.15 ppm (double bond), carbonyl acid in 180.03ppm and secondary alcohol in 79.69 ppm [Nick *et al.*, 1994; Mahato, Kundu, 1994].

The methylation EF-1 with diazomethane promoted the removal of hydrogen from the carboxyl acid and incorporation of a methyl group from the diazomethane, leading to formation of an ester, molecular weight 470. The derivatization and the formation of the ester are ideal possible to decrease the molecular interactions between the sample and a chromatographic column and thus decrease the retention time. An aliquot of esterified EF-1 (1.4mg) was subjected to MS electron impact (70 eV). The MS spectrum of esterified EF-1 confirmed the presence of a terpene class of lupanos due to the absence of peaks m/z 218 and m/z 203 characteristic of the series oleanane and ursane (rearrangement retro Diels-Alder ring C). The presence of the methyl ester group at C-28 is confirmed by the ion m/z 262 (10%). Other peaks were obtained m/z 208 (5%), m/z 190 (10%) and m/z 189 (100%) from the break ring C and the molecular ion m/z 470 5% [Budzikiewicz *et al.*, 1964]. The spectral data obtained from the EF-1 and the ester data were similar to those observed in the literature to betulinic acid [Nick *et al.*, 1994, Mahato, Kundie 1994; Budzikiewicz *et al.*, 1964].

After calibration with standard of betulinic acid, the monthly extracts from leaves of *Eugenia florida* were analyzed. Those extracts were analyzed in triplicate and the average areas corresponding to betulinic acid was calculated. From these average areas, percentage composition of the betulinic acid in the extract were calculated using the linear equation generated during calibration of betulinic acid (Figure 2) carried out in HPLC-UV and GC-FID (Table 1).

Month (year)	CG-FID (%)	HPLC (%)
August (2009)	7.43	8.01
September (2009)	16.83	8.57
October (2009)	26.27	11.18
November (2009)	17.97	8.36
December (2009)	8.24	4.83
January (2010)	6.90	2.79
February (2010)	7.92	5.12
March (2010)	23.03	6.12
April (2010)	9.68	6.31
May (2010)	14.85	9.16
June (2010)	19.33	9.99
Jully (2010)	15.23	5.62

Table 1. Quantification (w/w) of betulinic acid present in ethanol extracts from leaves *of Eugenia florida* determined by GC-FID and HPLC-UV at 210nm

4. Conclusions

Several activities are being attributed to betulinic acid, however, despite all of their potential pharmacological, it is still obtained by extraction of the bark and heartwood of some [Soler, 1996], synthetic processes [Evers et al., 1996] and by biotransformation [Galgon, 2005]. Unlike these traditional species whose income was less than 3%, we found that betulinic acid was present in all extracts analyzed (table 1), with yields well above those found in the literature.

The betulinic acid level in the *E. florida* leaves increased significantly in the May, June, Jully (autumn - winter) and, September, October and November (winter) which was mainly due to the accumulation of this compound in vegetal tissue. Some authors related with the pentacyclic triterpenes, just as betulinic, acid ursolic, acid, β-amyrine and lupeol, are supposed to be toxic to insects, due to their ability to inhibit acyl chain packing in the lipid bilayers of the insect membranes [Rodriguez *et al.*, 1997; Prades et al., 2011].

These fluctuations observed in the months described in Table 1 may be related to the chemical ecology of *Eugenia florida* as, for example, the attraction of pollinators or the reproductive phenology of the specimens

It is possible that the increased concentration of betulinic acid in the month of March is due to the large amount of rainfall characteristic of the Rio de Janeiro, state. However, more research is needed to determine whether other factors may be influencing the concentration of this metabolite, verify that specimens from other regions have the same or different behavior and examine whether the effect of the solvent can affect the increase in the concentration of this metabolite

Acknowledgements

The authors thank the (FAPERJ) for the financial support. We also acknowledge the Fundação Oswaldo Cruz, Farmanguinhos – PMA (Plataforma de Métodos Analíticos) for GC/FID, GC/MS analysis.

Author details

Alaíde S. Barreto[1,3], Gláucio D. Feliciano[1,4],
Cláudia Cristina Hastenreiter da Costa Nascimento[1], Carolina S. Luna[2], Bruno da Motta Lessa[2], Carine F. da Silveira[2], Leandro da S. Barbosa[2], Ana C. F. Amaral[2] and Antônio C. Siani[2]

1 Laboratory of Analysis Chemical - Biological (LAQB), Foundation of State University Center West Zone (UEZO), Brazil

2 Natural Products Laboratory, Oswaldo Cruz Foundation Farmanguinhos - Oswaldo Cruz Foundation (Fiocruz), Rio de Janeiro, Brazil

3 Department of Chemistry, University Severino Sombra (USS) Vassouras, Rio de Janeiro, Brazil

4 Universidade Estácio de Sá (UNESA), Área De Ciências Biológicas e da Saúde, Rio de Janeiro, Brazil

References

[1] Budzikiwicz, H, Djerassi, C, & Williams, D. H. (1964). Structure Elucidation of Natural Products by Mass Spectrometry. Volume II: Steroids, terpenoids, sugars and miscllaneous classes Holdey-day, INC, São Francisco, London, Amsterdam, 306p.

[2] Chatterjee P; Kouzi SA; Pezzuto JM; Hamann MT; ((2000). Biotransformation of the antimelanoma agent betulinic acid by Bacillus megaterium ATCC 13368. Applied and environmental microbiology. , 66(9), 3850-3855.

[3] Consolini, A. E, Baldini, O. A. N, & Amat, A. G. (1999). Pharmacological basis for the empirical use of *Eugenia uniflora L.* (Myrtaceae) as anthypertensive. Journal of Ethnopharmacology, , 66, 33-39.

[4] Evers, M, Poujade, C, Soler, F, Ribeil, Y, James, C, Lelièvre, Y, Geguen, J. C, Reisdorf, D, Morize, I, Pauwels, E, De Clercq, E, Hènin, Y, Bousseau, A, & Mayaux, J. F. Le Pecq, J.B.; Dereu N. ((1996). *Betulinic Acid Derivatives: A new class of HIV type 1 specific inhibitors with a new mode of action*. Journal of Medicinal of Chemistry , 39, 1056-68.

[5] Galgon, T, Wohlrab, W, & Drager, B. (2005). Betulinic acid induces apoptosis in skin cancer cells and differentiation in normal human keratinocytes. Experimental Dermatology. October, , 14(10), 736-743.

[6] Galgon, T, Höke, D, & Dräger, B. (1999). *Identification and Quantification of Betulinic Acid. Phytochem* Anal. 10. , 187-190.

[7] Gobbo-neto, L, & Lopes, N. P. (2007). Medicinal plants: factors of influence on the content of secondary metabolites. Química Nova. Mar./Apr. São Paulo., 30(2)

[8] Hendriks, H, Wildeboer, Y. A, Engels, G, Bos, R, & Woerdenbag, H. J. (1997). The content of parthenolide and its yield per plant during the growth of *Tanacetum parthenium.* Planta Medica , 63, 356-359.

[9] Junges, M. J, Fernandes, J. B, Vieira, P. C, Silva, M. F. G, & Filho, E. R. (1999). *The use of ^{13}C and ^{1}H-NMR in the structure elucidation of a new nor-lupane triterpene.* Journal Brazilian Chemical Society , 10(4), 317-320.

[10] Leonard, J, Lygo, B, & Procter, G. Advanced Pratical Organic Chemistry, 2 nd ed. Chapman & Hall, (1995).

[11] Lunardi, I, Peixoto, J. L. B, Silva, C. C, & Shuquel, I. T. A. Basso E.A.E; Vidotti G.J ((2001). Triterpenic Acids from Eugenia moraviana. Journal Brazilian Chemical Society, , 12, 180-183.

[12] Mahato, S. B, & Kundu, A. P. (1994). C NMR spectra of pentacyclic triterpenoids- A compilation and some salient features. Phytochemistry , 37, 1517-1575.

[13] Mayaux, J. F, Bousseaux, A, Panwels, R, De Clerq, E, & Pecq, J. B. (1994). *Triterpenes derivatives that block the entry of human immunodeficiency virus type I into cells.* Proceedings National Academy Sciences, , 91, 3564-3568.

[14] Mitscher, L. A, Pillai, S, & Shankel, D. M. (2000). Some transpacific thoughts on the regulatory need for standardization of herbal medical products. Journal Food and Drug Anal, n. 4, , 8, 229-234.

[15] Navarro, F. N, Souza, M. M, Neto, R. A, Golin, V, Niero, R, & Yunes, R. A. Delle Monache, F.; Cechinel Filho, Phytochemical analysis and analgesic properties of Curcuma Zedoaria grwn in Brasil. Phytomedicine, n.9, , 2002, 427-432.

[16] Oliveira, B. H, Santos, C. D. A, & Espíndola, A. P. D. M. (2002). *Determination of the triterpenoids, Betulinic Acid in Doliocarpus schottianus by HPLC*. Phytochemistry. Anal. , 13, 95-98.

[17] Oliveira, C. M, Moos, H. W, Chayer, P, & Kruk, J. W. *Variations in D/H and D/O from New Far Ultraviolet Spectroscopic Explorer Observations.* The Astrophysical Journal, (2006)., 642, 283-306.

[18] Pezzuto, J. M, Dasgupta, T. K, & Kim, D. S. H. L. United States Patent nº 5,869,535; February 9, (1999).

[19] Pisha, E, Chai, H, Lee, I. S, Chagwedera, T. E, Farnsworth, N. R, Cordell, G. A, Beecher, C. W. W, Fong, H. H. S, Kinghorn, A. D, Brown, D. M, Wani, M. C, Wall, M. E, Hieken, T. J, Dasgupta, T. K, & Pezzuto, J. M. (1995). *Discovery of betulinic acid as a selective inhibitor of human-melanoma that functions by induction of apoptosis.* Nature Medicine , 1, 1046-1051.

[20] Prades, J, Vögler, O, Alemany, R, & Gomez-florit, M. Funari, S.S; Ruiz-Gutiérrez, V.; Barceló, F. ((2011). Plant pentacyclic triterpenic acids as modulators of lipid membrane physical properties Biochimica et Biophysica Acta (BBA)- Biomenbranes. , 1808

[21] Robinson, T. Metabolism and function of alkaloids in plants. Science, (1974). , 184, 430-435.

[22] Rodriguez, S, Garda, H, Heinzen, H, & Moyna, P. (1997). Effect of plant monofunctional pentacyclic triterpenes on the dynamic and structural properties of dipalmitoylphosphatidylcholine bilayers Chemical Physical Lipids , 89-119.

[23] Siani, A. C, Sampaio, A. L. F, Sousa, M. C, Henriques, M. G. M. O, & Ramos, M. F. S. (2000). Óleos essenciais, potencial antiinflamatório. Revista Biotecnologia Ciência & Desenvolvimento, n 16, , 3, 38-43.

[24] Soler, F, Poujade, C, Evers, M, Carry, J. C, Hènin, Y, & Mayaux, J. F. Le Pecq J. B.; Dereu, N. ((1996). *Betulinic acid derivates: a new class of HIV type 1 entry.* J. Med. Chem. , 39, 1069-83.

[25] Spring, O, Bienert, U, & Klemt, V. (1987). Sesquiterpene lactones in glandular trichomes of sunflower leaves. Plant Physiol. , 130, 433-439.

Chapter 8

Natural Products from Semi–Mangrove Plants in China

Xiaopo Zhang

Additional information is available at the end of the chapter

http://dx.doi.org/10.5772/55933

1. Introduction

Mangroves are various kinds of trees up to medium height and shrubs that grow in saline coastal sediment habitats in the tropics and subtropics-mainly between latitudes 25° N and 25° S [1]. Mangrove plants were comprised of true-mangrove plants and semi-mangrove plants. The true-mangrove plants were woody plants, which only grew in the intertidal zone and couldn't survive in the land. Semi-mangroves were woody plants that could both grow in the intertidal zone and in the land. The differences between the two kinds of mangroves were the specificity of living habitats of the true-mangrove and the amphibiotic living habitats of the semi-mangrove as shown in Figure 1. Meanwhile, all of them were woody plants and could grow in the specific environment of intertidal zone, and the latter was the basis of the diversified chemical constituents and biological activities of the mangroves [2].

True-mangroves   Semi-mangroves

Figure 1. Pictures of some true and semi-mangroves

The word's mangrove plants have 84 species (including 12 varieties) in 24 genera and 16 families. Of which, true mangrove plants have 70 species (including 12 varieties) in 16 genera and 11 families, and semi-mangrove plants 14 species in eight genera and five families [3]. 12 species of semi-mangroves were grown in China including *Barringtonia racemosa, Cerbera manghas, Dolichandrone spathacea, Pluchea indic, Hernandia nymphiifolia, Pongamia pinnata, Pemphis acidula, Hibiscus tiliaceus, Thespesia populnea, Premna obtusifolia, Clerodendrum inerme, Heritiera littoralis* [3]. Hainan province of China is most rich in semi-mangroves, where all of the 12 species were spread there. 8 species were grown in Guangxi province, 10 species were distributed in Taiwan province, 7 species were found in Fujian province, 7 species were grown in Hongkong, and 3 species were found in Macao [4].

Most semi-mangroves possessed medicinal usages were utilized as folk medicine in many provinces of China. For example, the seeds of *Cerbera manghas* were used as emetic and purgative medicine in Hainan province [1]. The leaves and branches of *Hibiscus tiliaceus* were used as the agents of clearing heat and reducing the swelling. The crude extract of *Pongamia pinnata* can effectively inhibit pathogen of the multiple evanescent white dot syndromes, and reduce mortality. The seeds of *Barringtonia racemosa* showed anti-cancer and anti-microbial activities. *Thespesia populnea* possessing relieving pain, anti-inflammatory, anti-microbial, and antioxidant activities can also protect the liver damage induced by carbon tetrachloride.

Chemical constituents isolated form semi-mangrove plants with various and unique structures including flavanoids, lignans, sesquiterpenoids, diterpenoids, triterpenoids, steroids, and alkaloids et al. For example, Abe and coworkers obtained cardiac glycosides from the seeds of *Cerbera manghas* [4]. Alis isolated some oxygenated sesquiterpenoids from the roots of *Hibiscus tiliaceus*. Some flavanoids as glabone were isolated from *Pongamia pinnata* and Wang obtained 5 thiophene derivatives from *Pluchea indic*.

Meanwhile, the biological activities of the isolated compounds were studied extensively. Some were found to have obvious biological activities. For example, cardiac glycosides showed obvious anti-cancer activity [11] and bartogenic acid can inhibit the activity of α-glycosidase and amylase [13].

In short, semi-mangrove plants played an important role in curing disease and finding new chemical entities. We highlight that it will become more and more significant to the research and development of new drugs. In this review, chemical constituents and biological activities of semi-mangrove plants were mainly regarded and they are as follows.

2. Chemical constituents

2.1. Flavanoids

Pongamia pinnata were rich in flavanoids compared with other semi-mangroves. Up to now, more than 50 flavaonoids have been isolated from *Pongamia pinnata* with the structural characteristic of furan or pyran ring parallelized with the skeleton of flavaonoids. Partial flavaonoids including flavone, flavonone, chalcone, dihydrochalcone were listed in Figure 3.

Figure 2. Pictures of 12 semi-mangroves spread in China

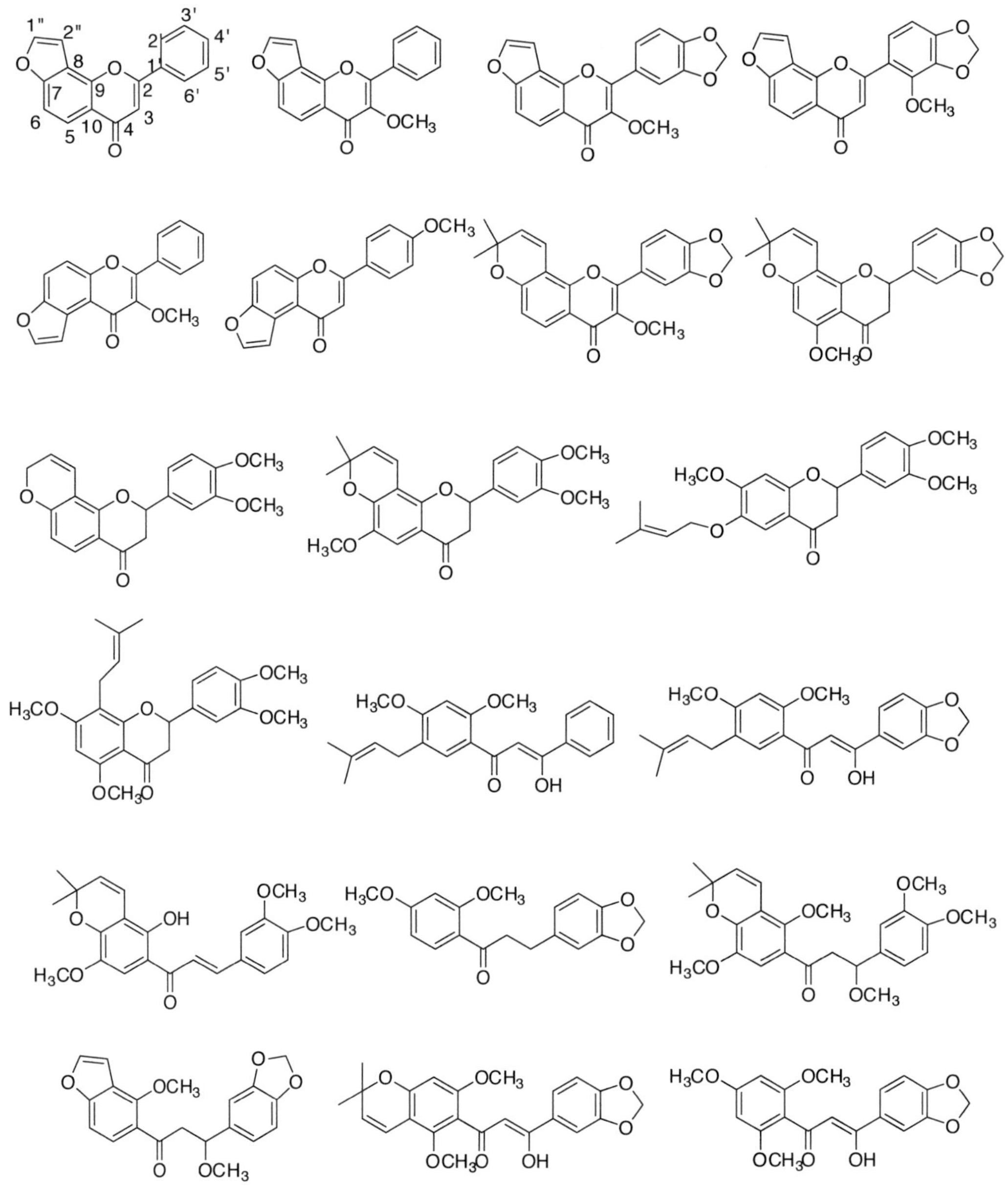

Figure 3. Flavonoids from *Pongamia pinnata*

2.2. Sesquiterpenoids

Sesquiterpenoids isolated form *Pluchea indic*, *Hibiscus tiliaceus*, and *Thespesia populnea* were given full attention. Most sesquiterpenoids isolated form *Pluchea indic* were eudesmane and eremophilane diterpenoid skeleton, For example, the compounds were depicted in Figure 4.

http://dx.doi.org/10.5772/55933

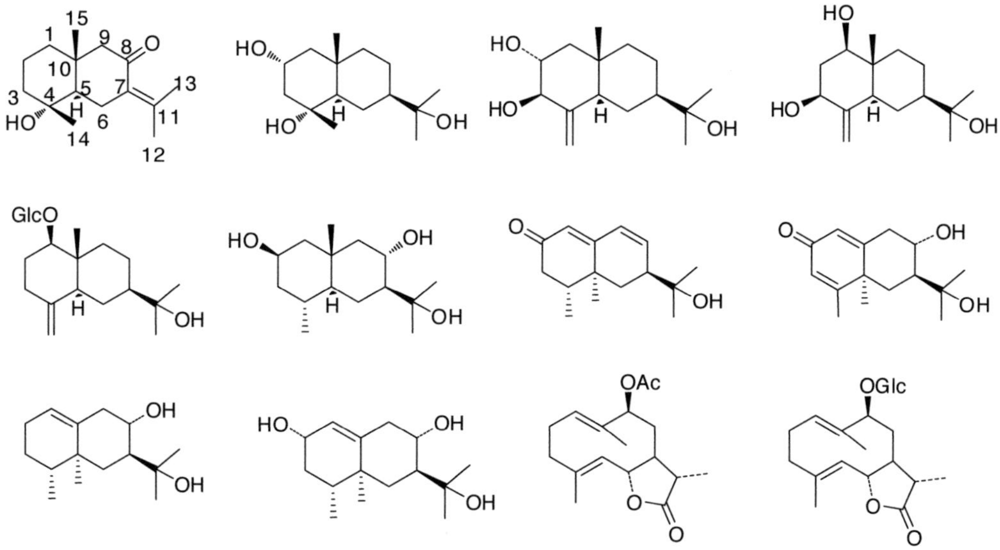

Figure 4. Structures of sesquiterpenoids from *Pluchea indica*

Sesquiterpenoids isolated form *Hibiscus tiliaceus* and *Thespesia populnea* with highly oxygenated structures attracted many scientists' attention. Nine oxygenated sesquiterpenoids were isolated from the heartwood of *Hibiscus tiliaceus* and they are shown in Figure 5.

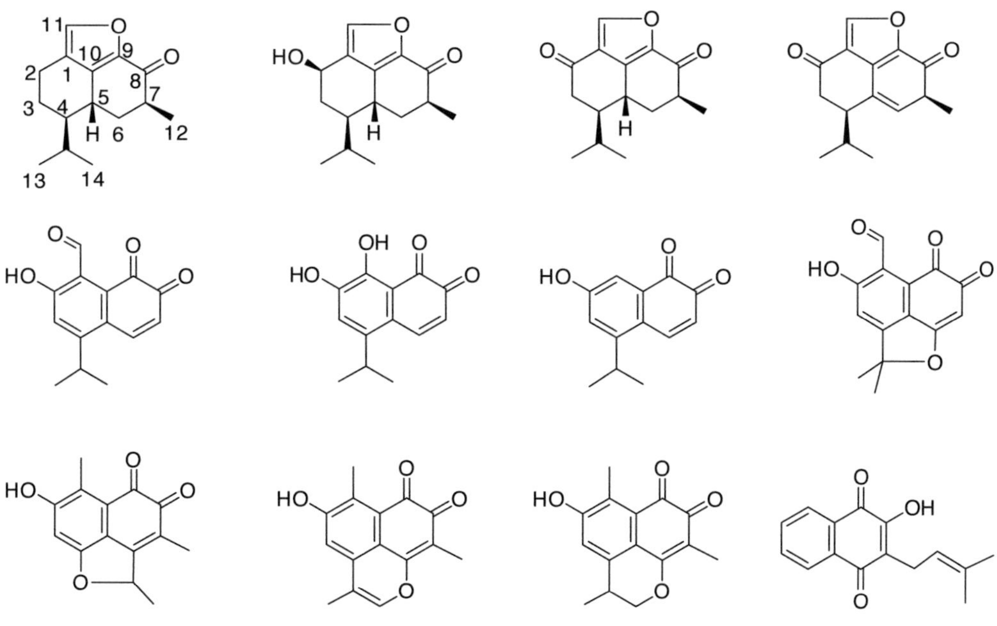

Figure 5. Structures of mansonones from *Hibiscus tiliaceus*

2.3. Triterpenoids

Some oleanane type triterpenoids with highly oxygenated were isolated form *Barringtonia racemosa* as shown in Fig 6. Six friedelane type triterpenoids were isolated from bark of *Hibiscus tiliaceus* collected from Hainan province as shown in Figure 7.

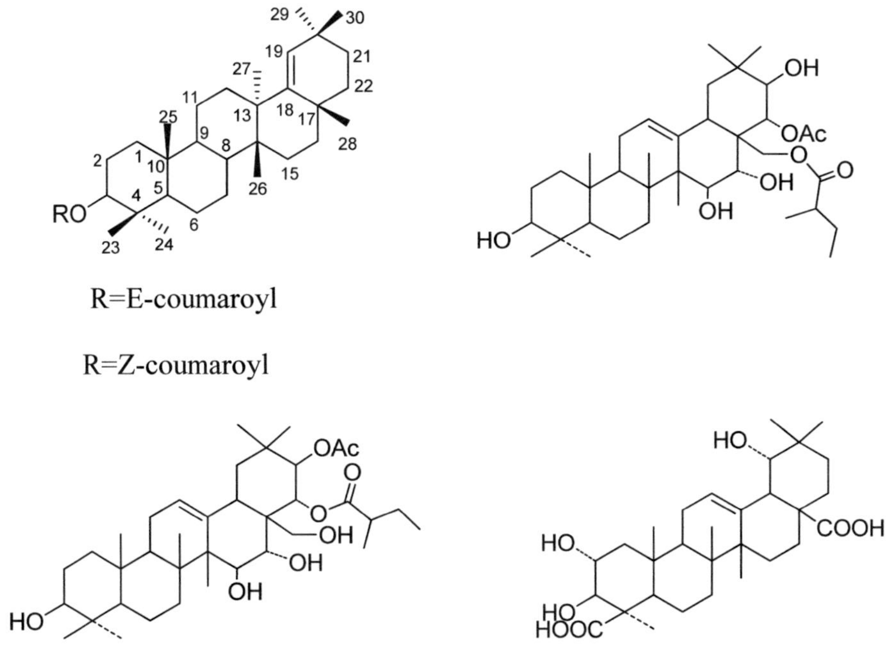

Figure 6. Triterpenoids from *Barringtonia racemosa*

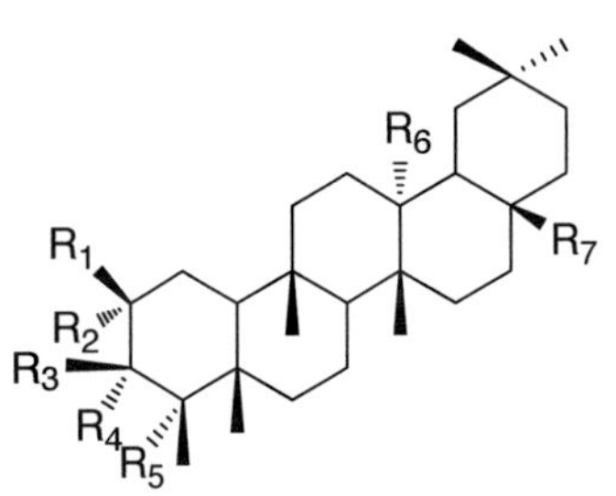

R_1=R_2=O, R_3=R_5=H, R_4=OH, R_6=R_7=CH_3 (1)
R_1=R_2=H, R_3=R_4=O, R_5=OH, R_6=R_7=CH_3 (2)
R_1=R_2=R_5=H, R_3=R_4=O, R_6=R_7=CH_3 (3)
R_1=R_2=R_4=R_5=H, R_3=OH, R_6=R_7=CH_3 (4)
R_1=R_4=R_5=H, R_2=R_3=OH, R_6=R_7=CH_3 (5)
R_1=R_2=R_5=H, R_3=R_4=O, R_6=R_7=COOH (6)

Figure 7. Structures of friedelane type triterpenoids from *Hibiscus tiliaceus*

2.4. Cardiac glycosides

More than 30 cardiac glycosides were isolated from the seeds of *Cerbera manghas* and the skeleton of the obtained compounds were calssified into three classess including digitoxigenin (A), Oleandrin (B), and tanghinin (C) as shown in Fig 8.

http://dx.doi.org/10.5772/55933

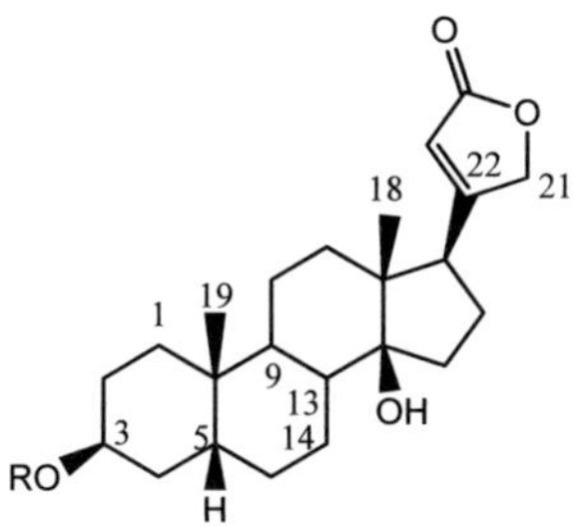

A

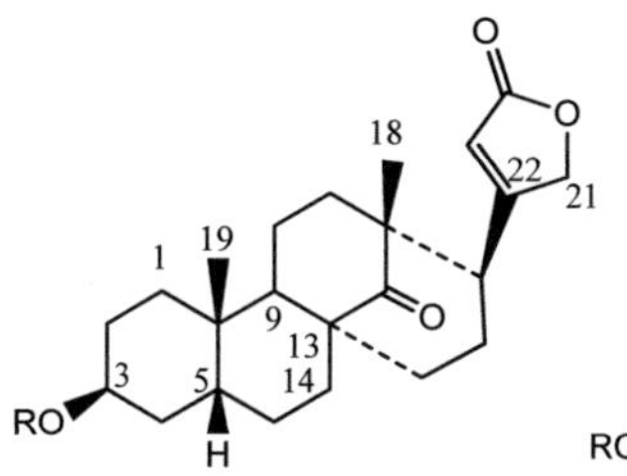

B

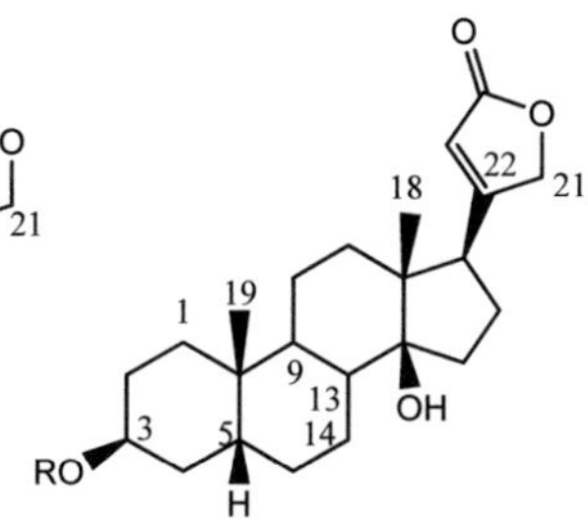

C

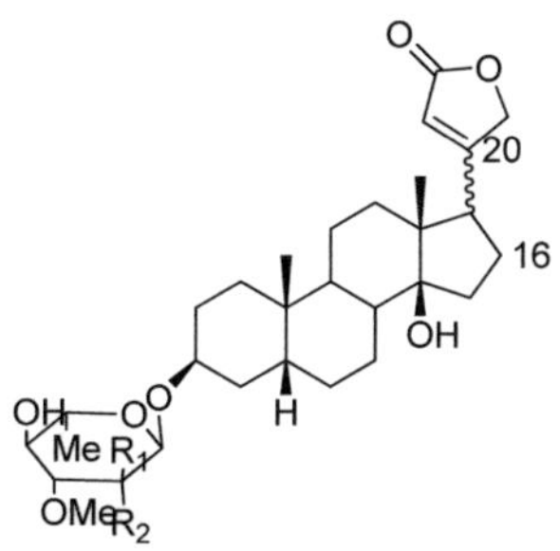

17β R1=OH;R2=H neriiforlin
17α R1=OH;R2=H 17α neriiforlin
17β R1=H;R2=OH
17α R1=H;R2=OH

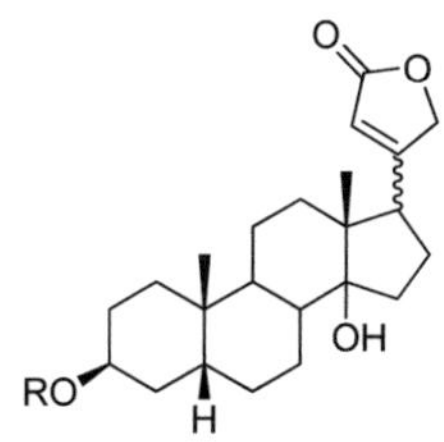

17β ,R=a
17β ,R=c
17β ,R=b
17α ,R=b
17β ,R=d

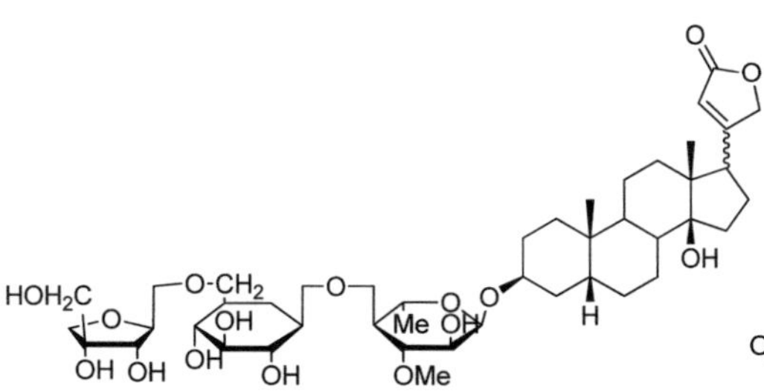

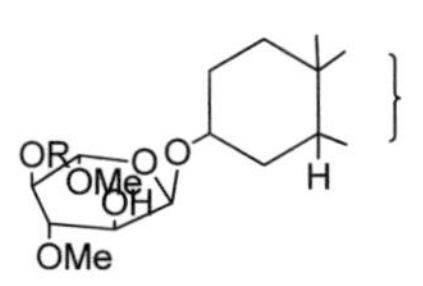

4
R=-Glc← Glc
6
R=-Glc← Glc

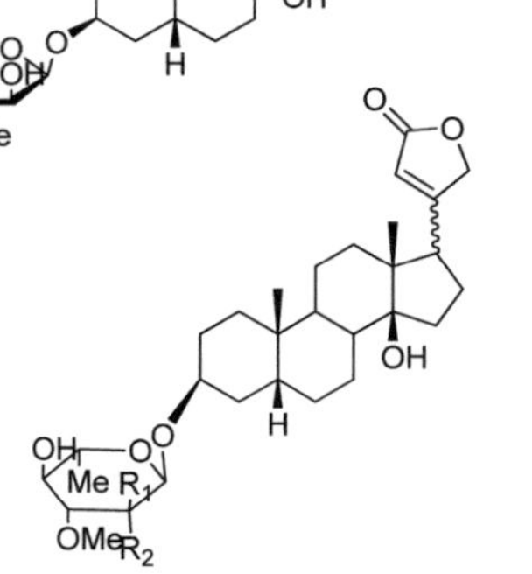

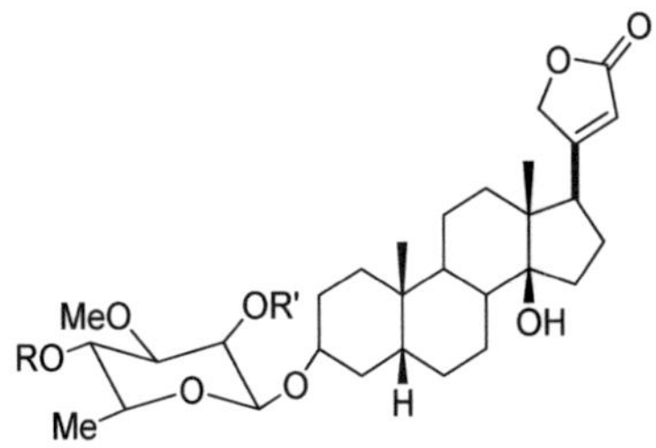

thevetin B R=β-gentiobiosyl R'=H
2'-O-Ac-thevetin B R=β-gentiobiosyl R'=OAc
cerberin R=H R'=OAc

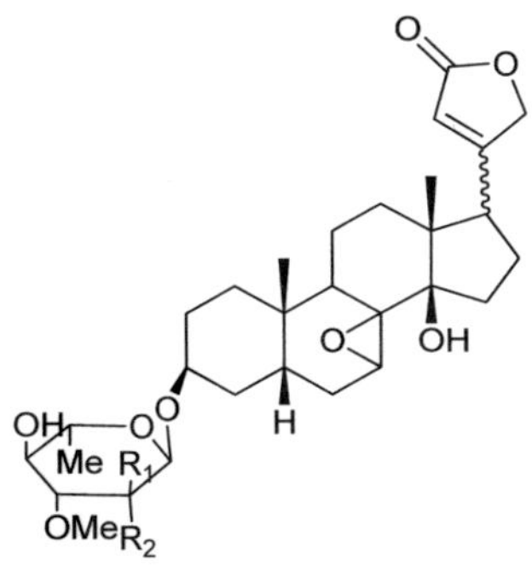

17β ;R1=OH;R2=H deacetyltanghinin
17α ;R1=H;R2=OH 17α-deacetyltanghinin
17β ;R1=H;R2=OH

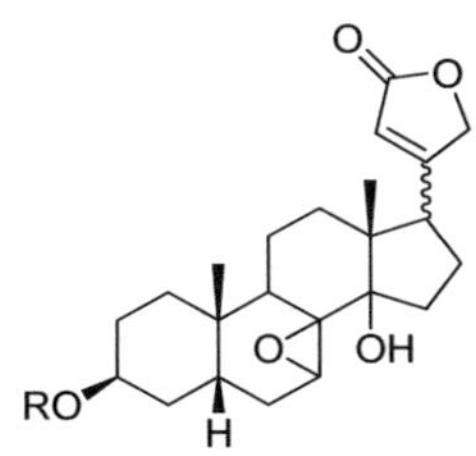

17α ,R=a
17β ,R=b
17α ,R=b
17β ,R=d

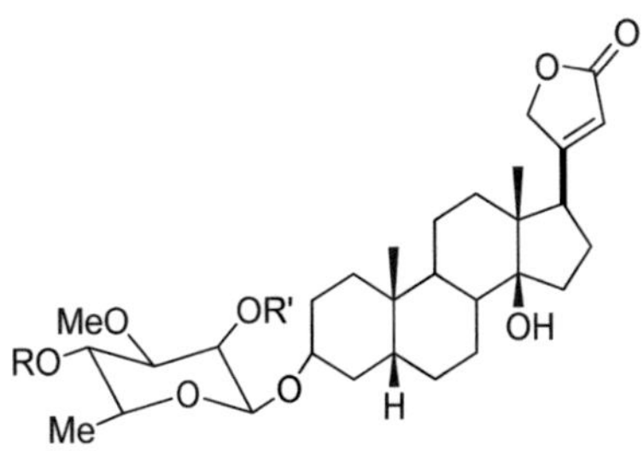

R=β-gentiobiosyl R'=OAc $C_{7,8}$β β-epxoy
R=β-gentiobiosyl R'=H $C_{7,8}$β β-epxoy

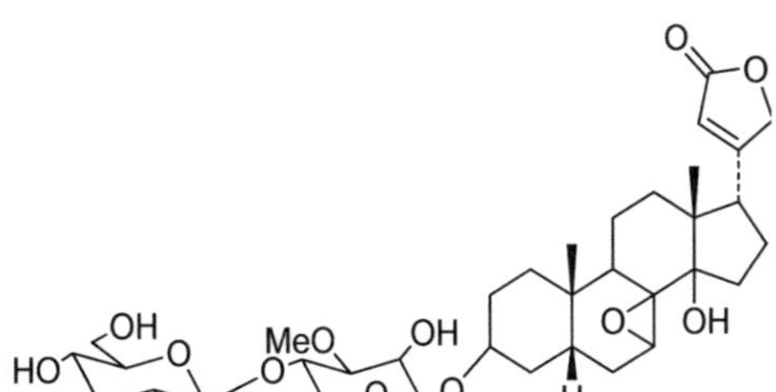

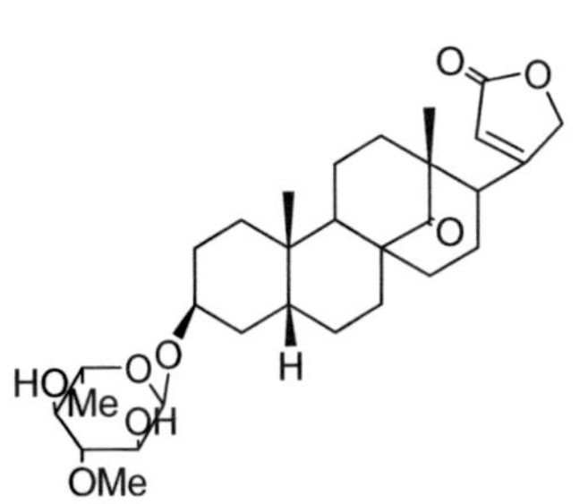

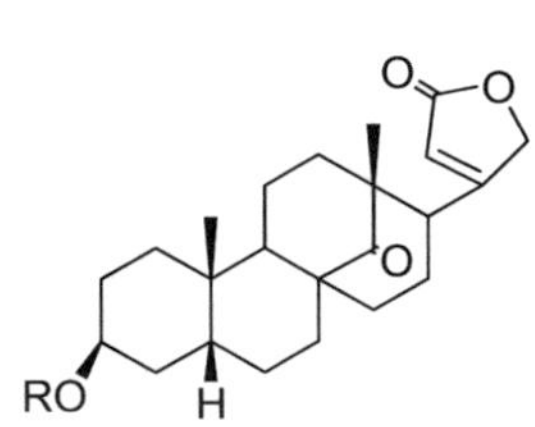

17α ,R=b

http://dx.doi.org/10.5772/55933

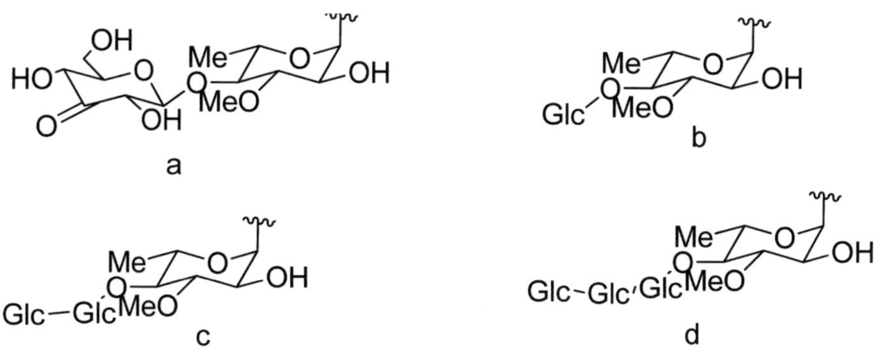

Figure 8. Cardenolides from *Cerbera manghas*

2.5. Lignans

Many lignans were isolated from the semi-mangroves, and lignans obtained from *Cerbera manghas* with unique structures have attracted more attention. These lignans were classified to be monomerlignans (1-4), sesquilignans (5-7), dilignans (8-16), sesterlignans (17) and trilignans (19). They are listed as belows.

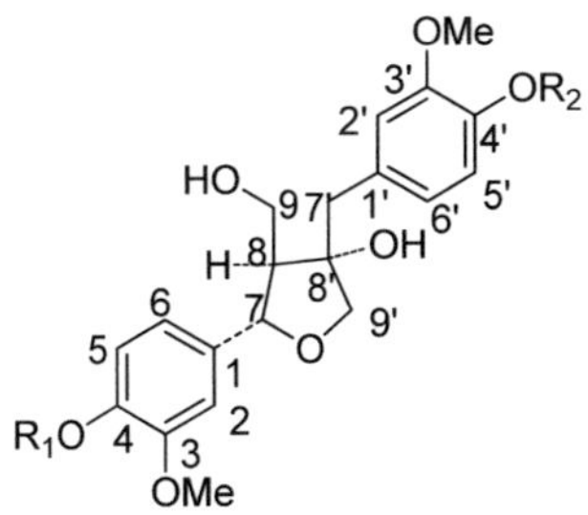

R_1=R_2=H

R_1=glucosyl,R_2=H

R_1=H,R_2=glucosyl

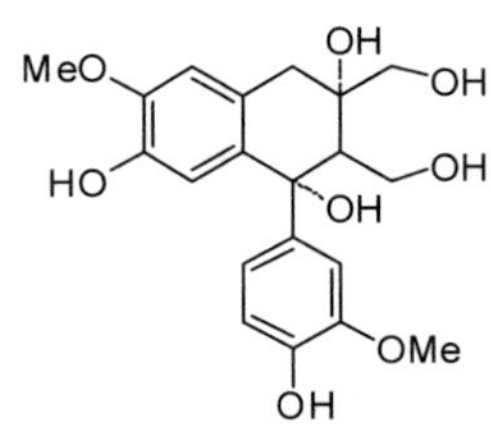

cycloolivil

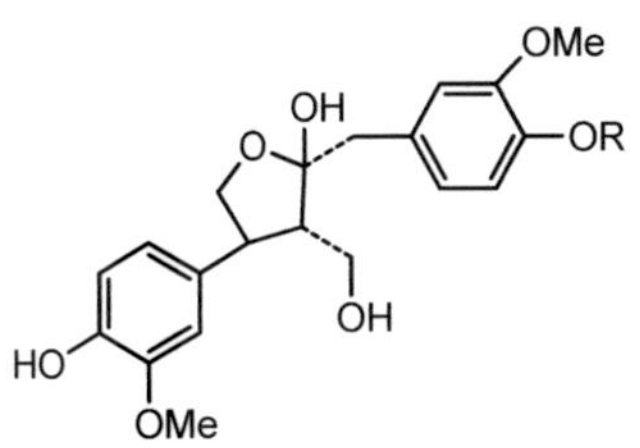

R=X(erythro) cerberalignan F

R=X(threo) cerberalignan G

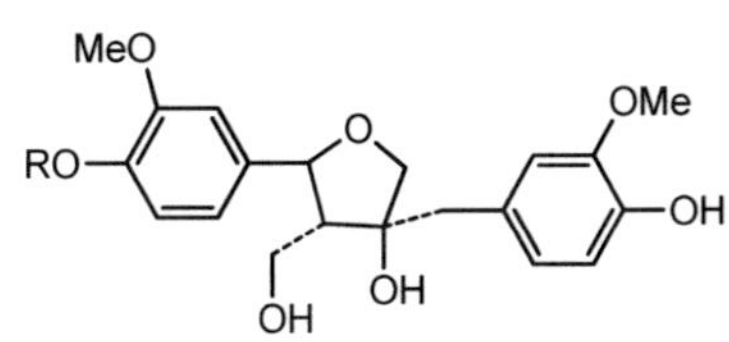

R=X(threo)

cerberalignan J

X=

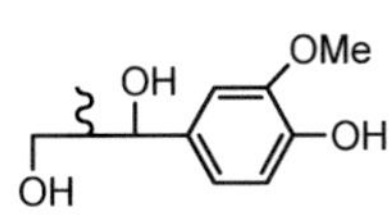

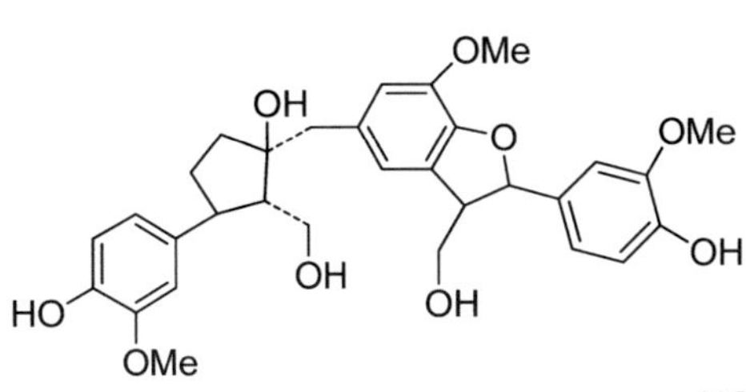

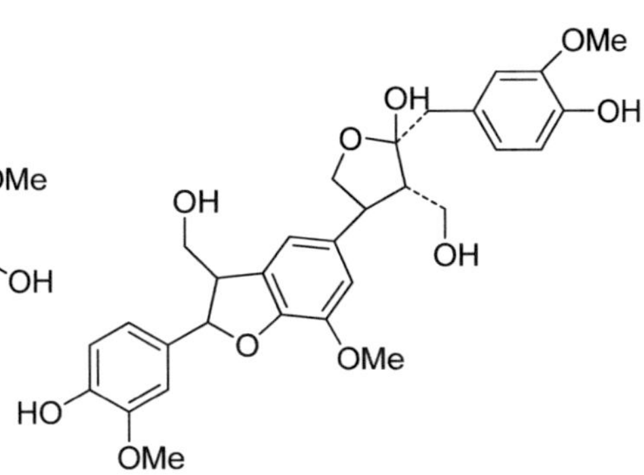

http://dx.doi.org/10.5772/55933

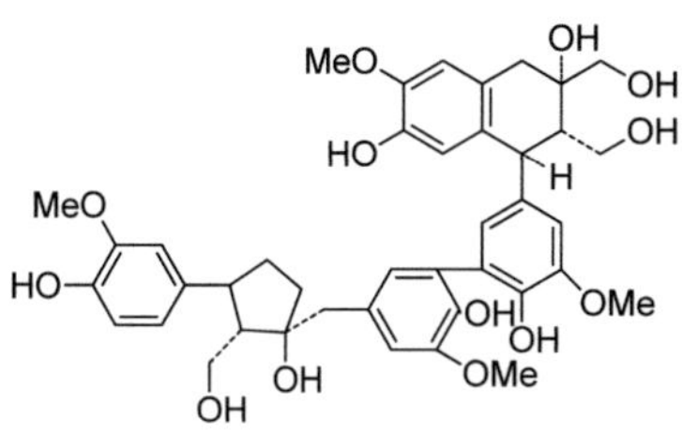

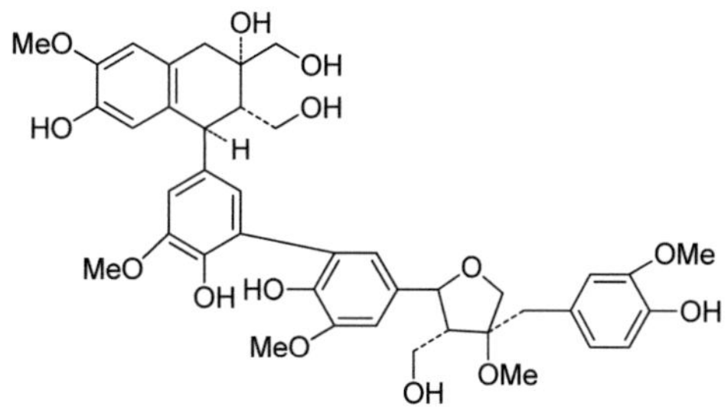

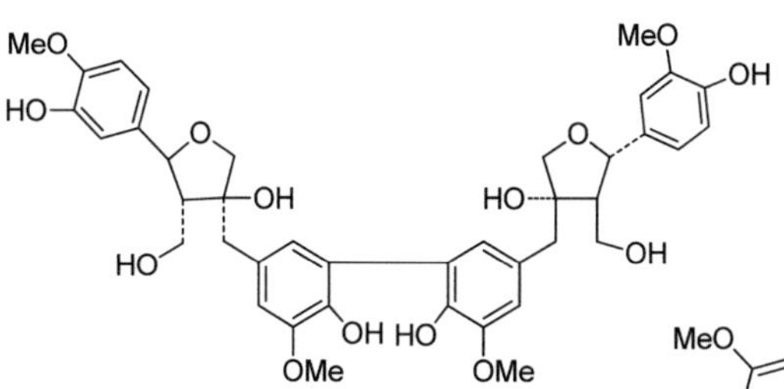

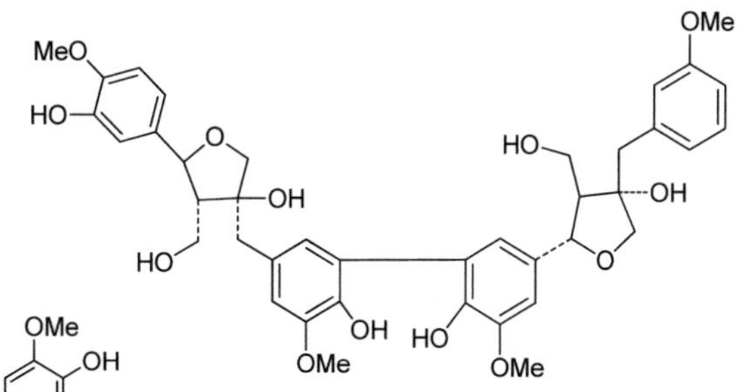

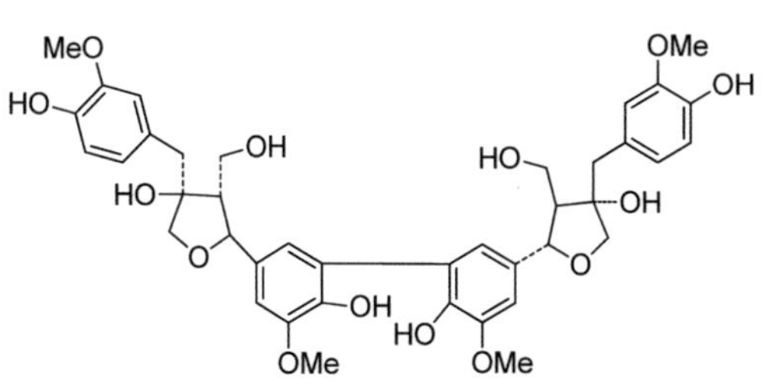

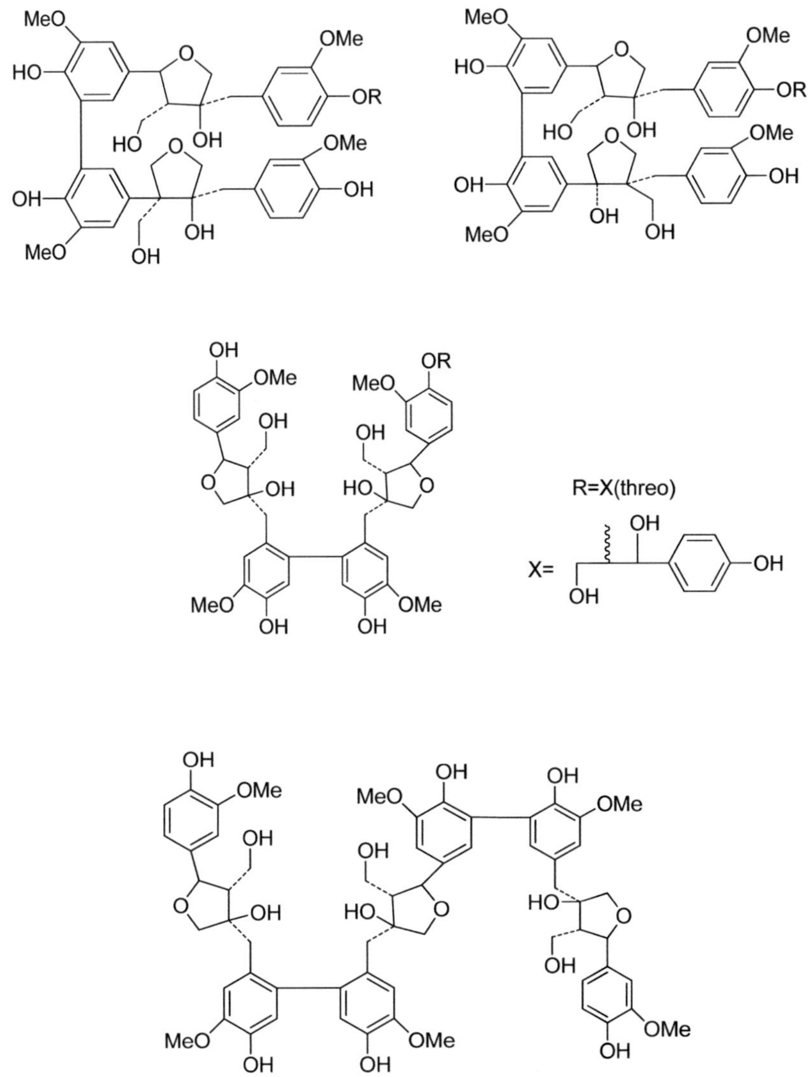

Figure 9. Structures of lignans from *Cerbera manghas*

2.6. Others

Apart from these above mentioned compounds, courmains, iridoids, and alkaloids have also been obtained from semi-mangroves distributed in China. For example, cerbinal, p-hydroxy-benzaldehyde, benzamide, n-hexadecane acid monoglyceride, loliolide, β-sitosterol, *et al.*

3. Biological activities

3.1. Antitumor activities

The seeds extract of *Cerbera manghas* were found possessing obvious cytotoxic activity against some human cancer cell lines by MTT methods. Feng and coworkers obtained two cardiac glycosides named GHSC-73 and GHSC-74. Further study suggested GHSC-73 and GHSC-74 can significantly inhibited growth and proliferation of HepG2 cells in dose-dependent manner. GHSC-73 inhibited the growth and proliferation of HepG2 cells through blocking S phase and inducing apoptosis, while GHSC-74 through blocking S and G2 phases and inducing apoptosis.Wang and coworkers tested 24 compounds isolated from *Pluchea indic* and found that valenc-l (10)-ene-8α, 13-diol showed inhibiting activity against Bel-7402 and A2780 cells. Lanceolatin B purified from *Pongamia pinnata* can prevent the development of cancer [20]. The cytotoxic activities of mansonone D, mansonone H, thespesone, and thespone were tested against MCF-7 cells by Johnson using the MTT methods. The results indicated that they all showed certain cytotoxic activities [21]. Ethnomedical survey has shown that the seeds of *Barringtonia racemosa* are traditionally used in certain remote villages of Kerala (India) to treat cancer like diseases. Thomas [22] tested the seed extracts for their anti-tumor activity and toxicity. Intraperitoneal (i.p.) daily administration of 50% methanol extract of this seed to mice challenged with 1 million Dalton's Lymphoma Ascitic (DLA) cells resulted in remarkable dose dependent anti-DLA activity in mice. The optimum dose was found to be 6 mg/kg. This dose protected all the animals challenged with the tumor cells. The efficacy of the drug was found to be better than that of a standard drug, vincristine in this tumor model. However, the oral administration showed only marginal activity compared to i.p. administration. The extract was found to be devoid of conspicuous acute and short-term toxicity to mice, when administered daily, (i.p.) for 14 days up to a dose of 12 mg/kg (which was double the concentration of optimum therapeutic dose). The treated mice showed conspicuous toxic symptoms only at 24 mg/kg. The LD (50) to male mice for a single i.p. dose was found to be 36 mg/kg. Consequently, they found that the seed extract is an attractive material for further studies leading to drug development. Anbu and coworkers [23] evaluated anti-tumor activity of the roots of *Hibiscus tiliaceus* against Dalton's Ascitic Lymphoma (DAL) in Swiss albino mice. A significant enhancement of mean survival time (MST) of *H. tiliaceus* treated tumor bearing mice was found with respect to control group. *H. tiliaceus* treatment was found to enhance peritoneal cell counts. When these *H. tiliaceus* treated animals under-went intraperitoneal (i.p.) inoculation with DAL cells, tumor cell growth was found to be inhibited. The results indicated that, *H. tiliaceus* treated group were able to reverse the haematological parameters, protein and Packed

Cell Volume (PCV) consequent to tumor inoculation with in fourteen days after the transplantation.

3.2. Antibacterial activity

Khan and coworkers [24] used disc diffusion methods to test antibacterial activity of the ethanol extract of *Barringtonia racemosa* roots, its chloroform soluble fraction, and a there from an isolated clerodane diterpenoid (Nasimalun A). The results presented that they all showed potent activity in inhibiting the growth of 19 strains of bacterial with the ethanol extract as the most activity part. A marble cup method was used by Goyal [25] to test the antimicrobial activities of the crude methanolic extract of *Barringtonia asiatica* (leaves, fruits, seeds, stem and root barks) and the fractions (petrol, dichloromethane, ethyl acetate, and butanol) and all the extract exhibited a very good level of broad spectrum antibacterial activity. Baswa [26] evaluated the antibacterial activity of *Pongamia pinnata* seed oil *in vitro* against fourteen strains of pathogenic bacteria. Using the tube dilution technique, it was observed that 57.14 and 21.42% of the pathogens were inhibited at 500 μl/ml, 14.28 and 71.42% at 125 μl/ml, and 28.57 and 7.14% at 250 μl/ml of *Pongamia pinnata* oils. The activity with both the oils was bactericidal and independent of temperature and energy. Most of the pathogens were killed more rapidly at 4°C than 37°C. The activity was mainly due to the inhibition of cell-membrane synthesis in the bacteria.

3.3. Anti–inflammatory analgesic activity

Srinivasan and coworkers studied the anti-inflammatory activity of 70% ethanolic extract of *Pongamia pinnata* leaves (PLE) in acute, subacute and chronic models of inflammation in rats. *Per os* (p.o.) administration of PLE (300, 1000 mg/kg) exhibited significant anti-inflammatory activity in acute (carrageenin, histamine, 5-hydroxytryptamine and prostaglandin E_2-induced hind paw edema), subcute (kaolin-carrageenin and formaldehyde-induced hind paw edema) and chronic (cotton pellet granuloma) models of inflammation. These results indicate that PLE possesses significant anti-inflammatory activity without ulcerogenic activity suggesting its potential as an anti-inflammatory agent for use in the treatment of various inflammatory diseases. The antinociceptive activity of a 70% ethanol extract of *Pongamia pinnata* leaves (PLE) was also investigated by Srinivasan [28] in different models of pain in mice and rats. Per os (p.o.) administration of the PLE (100-1000 mg/kg) produced significant antinociceptive activity in the hotplate and tail flick (central) as well as in acetic acid writhing and Randall-Selitto (peripheral) nociceptive tests. Narender[29] evaluated the antinociceptive and anti-inflammatory activities of different extracts of *Hibiscus tiliaceus* (Malvaceae). The antinociceptive investigations were carried out against two types of noxious stimuli, chemical (acetic acid-induced writhing) and thermal (hotplate and tail immersion tests). The different leaves extracts of *Hibiscus tiliaceus* (250 and 500 mg/kg, orally) possessed a significant anti-inflammatory activity on carragennan-induced paw edema in rat at the second and third hour. All the extracts significantly inhibited the acetic acid induced abdominal contractions in mice in order methanolic>chloroform>petroleum ether extract. The extracts showed the significant antinociceptive activity at dose of 250 mg/kg and 500 mg/kg ($p<0.01$) at 60 min after extracts administration.

3.4. Others

Chakraborty reported the bark, seeds of *Barringtonia acutangula* could be used as a fish poison. Pongamia pinnata was evaluated by Elanchezhiya [31] for antiviral properties against herpes simplex virus type-1 (HSV-1) and type-2 (HSV-2) by *in vitro* studies in Vero cells. A crude aqueous seed extract of *P. pinnata* completely inhibited the growth of HSV-1 and HSV-2 at concentrations of 1 and 20 mg/ml (w/v), respectively.

4. Conclusions

Among the 12 mangroves, *Pongamia pinnata, Cerbera manghas, Barringtonia racemosa* have been carried on deep researches concerning their chemical constituents, biological activities. Some compounds isolated from the semi-mangroves were proved to have obvious activities as the two cardenolides of GHSC-73 and GHSC-74. However, as proposed by Shao Changlun "Some semi-mangrove plants have not yet been analyzed and interpreted on their chemical constituents and pharmacological effects", the further researches of the semi-mangrove plants will make an important contribution to for the finding of new drugs.

Acknowledgements

We would like to express our sincerest gratitude to the financial support of initial funding by Hainan Medical University (Project No. HY2010-012). We also thank Dr. Yinfeng Tan for helpful discussions regarding the pharmacological analysis.

Author details

Xiaopo Zhang

School of Pharmaceutical Science, Hainan Medical University, Haikou, PRC

References

[1] Lin, P, & Fu, Q. Environmental ecology and economic utilization of mangroves. Beijing: higher education press, (1995).

[2] Lin, P. Zoology of China's mangroves. Beijing: Science press, (1997).

[3] Wang, W. Q, & Wang, M. Mangroves in China. Beijing: Science press, (2007).

[4] Abe, F, & Yamauchi, T. Studies on Cerbera. I. Cardiac glycosides in the seeds, bark, and leaves of *Cerbera manghas*. *Chemical Pharmaceutical Bulletin*, (1997).

[5] Ali, S, Singh, P, & Thomson, R H. Naturally occurring quinones.Part 28.Sesquiterpenoid quinones and related compounds from *Hibiscus tiliaceus*.*J Chem Soc Perkin Trans* l. (1980). , 1980, 257-259.

[6] Das, P, Ganguly, A, & Guha, A. Glabone, a new furanoflavone from *Pongamia glabra* [J]. *Phy tochemistry*, (1987).

[7] Tanaka, T, Iinuma, M, & Yuk, i K. Flavonoids in root bark of *Pongamia pinnata* [J]. *Phytochemistry*, (1992).

[8] Tian, Y, Wu, J, & Zhang, S. Flavonoids from leaves of *Heritiera littoralis*. *Journal of Chinese Pharmaceutical Science*, (2004).

[9] Wang, J. Chemical constituent investigation of Mangrove Plant *Pluchea Indica* (L.) [D]. Master's thesis of Shenyang Pharmaceutical University, (2008).

[10] Rameshthangam, P, & Ramasamy, P. Antiviral activity of bis (methylheptyl) phthalate isolated from *Pongamia pinnata* leaves against white spot syndrome virus of *Penaeus monodon fabricius*. *Virus Research*, (2007).

[11] Guo, Y. W. The preparation and application of a sort of cardiac glycoside derivatives [(CN1715292), 2006.

[12] Feng, C. Chemical constituents of *Hibiscus tiliaceus* and *Thespesia populnea* [D]. Master's thesis of Chinese Academy of Sciences, (2008).

[13] Gowri, P. M, Tiwari, A. K, Ali, A. Z, & Rao, J. M. Inhibition of alpha-glucosidase and amylase by bartogenic acid isolated from *Barringtonia racemosa* Roxb. seeds. *Phytotheraphy Research*. (2007).

[14] Huang, B. X. Chemical constituents and antioxidant of *Pongamia Pinnata* [D]. Master thesis of Guangxi Medical University, (2002).

[15] Tian, Y, Wu, J, & Zhang, S. Advances in research of chemical constituents and pharmacological activities of semi-mangrove medicinal plant *Thespesia populnea* [J]. *Chinese Traditional and Herbal Drugs*, (2003).

[16] Gowri, P. M, Radhakrishnan, S. V, Basha, S. J, Sarma, A. V, & Rao, J. M. Oleanane-type isomeric triterpenoids from *Barringtonia racemosa*. *Journal Natural Product*. (2009).

[17] Yang, Y, Deng, Z, Proksch, P, & Lin, W. Two new en-oleane derivatives from marine mangrove plant, *Barringtonia racemosa*. *Pharmazie*, (2006). , 18.

[18] Sun, H. Y, Long, L. J, & Wu, J. Chemical constituents of mangrove plant *Barringtonia racemosa*. *Journal of Chinese Medicinal Materials*, (2006).

[19] Feng, B. Effects of extracts from Seeds of *Cerbera Manghas* on Cell proliferation, cell Cycle progression and apoptosis of human hepatocellular carcinoma HepG2 Cells

and their Mechanisms [D]. Master thesis of Secondary Military Medical University, (2009).

[20] Chang, L C, & Gerhauser, C. Activity guided isolation of constituents of *Tephrosia purpurea* with the potential to induce the phase II enzyme, quinone, reductase [J]. *Journal Nauralt Product*, (1997).

[21] Johnson, I J, Gandhidasan, R, & Murugesan, R. Cytotoxicity and superoxide anion generation by some naturally occurring quinones [J]. >*Free Radical Biology & Medicine*, (1999).

[22] Thomas, T. J, Panikkar, B, & Subramoniam, A. Antitumor property and toxicity of *Barringtonia racemosa* Roxb seed extract in mice. *J Ethnopharmacology*, (2002).

[23] Anbu, J S, Syam, M, et al. Antitumour activity of *Hibiscus tiliaceus* Linn. Roots. *Iranian Journal of Pharmacology&Therapeutics*, (2008).

[24] Khan, S, Jabbar, A, & Hasan, C. M. Antibacterial activity of *Barringtonia racemosa. Fitoterapia,* (2001).

[25] Khan, M. R, & Omoloso, A. D. Antibacterial, antifungal activities of *Barringtonia asiatica. Fitoterapia,* (2002).

[26] Baswa M, Rath C C, Dash S K, et al. Antibacterial activity of karanj (*Pongamia pinnata*) and Iveem (*Azadirachta indica*) seed oil: a preliminary report [J]. *Microbios*, 2001, 105 (412): 183-189.

[27] Srinivasan, K, Uruganandan, M, & Al, S, L. J, *et al*. Evaluation of anti-inflammatory activity of *Pongamia p innata* leaves in rats [J]. *J Ethnopharmacology*, (2001).

[28] Srinivasan, K, Uruganandan, M, & Al, S, L. J, *et al*. Antinociceptive and antipyretic activities of *Pongamia pinnata* leave [J]. *Phytotherapy Research*, (2003).

[29] NarenderSunil Kumar, et al. Antinociceptive and Anti-Inflammatory Activity of *Hibiscus tiliaceus* Leaves. *International Journal of Pharmacognosy and Phytochemical Research*, (2009).

[30] Chakraborty, D. P, Nandy, A. C, & Philipose, M. T. *Barringtonia acutangula* (L.) Gaertn as a fish poison. *Indian Journal of Experimental Biology*, (1972).

[31] Elanchezh iyanMRajarajan S, Rajendran P, *et a*l. Antiviralproperties of the seed extract of an Indian medicinal plant: *Pongaia pinnata* against herpes simplex virues: *in vitro* studies on vero cells [J]. *Journal of Medical Microbiology*, (1993).